NEC3 Construction Contracts: 100 Questions and Answers

Which member of the NEC3 family of contracts should I use?

How do I choose and use my main and secondary options?

What are the roles and responsibilities of the various parties?

How should I effectively manage early warnings and compensation events?

Important questions can arise when working with NEC3 contracts; some of them have simple answers and others require more a detailed response. Whether you are an NEC3 beginner or an expert, the 100 questions and answers in this book are a priceless reference to have at your fingertips.

Covering issues that can arise from the full range of NEC3 forms, Kelvin Hughes draws on questions he has been asked during his 20 years working with NEC and presenting training courses to advise, warn of common mistakes, and explain in plain English how these contracts are meant to be used.

Kelvin Hughes spent 18 years in commercial management with major contractors, then the past 23 years as a consultant, including a four-year senior lectureship at the University of Glamorgan. He has been a leading authority on the NEC since 1996, was Secretary of the NEC Users' Group for ten years and has run over 1,500 NEC training courses.

NEC3 Construction Contracts: 100 Questions and Answers

Kelvin Hughes

LONDON AND NEW YORK

First published 2016
by Routledge
2 Park Square, Milton Park, Abingdon, Oxon OX14 4RN

and by Routledge
711 Third Avenue, New York, NY 10017

Routledge is an imprint of the Taylor & Francis Group, an informa business

British Library Cataloguing-in-Publication Data
A catalogue record for this book is available from the British Library

Library of Congress Cataloging in Publication Data
Names: Hughes, Kelvin (Engineering consultant)
Title: NEC3 construction contracts : 100 questions and answers / Kelvin Hughes.
Other titles: New engineering contracts third construction contracts
Description: Abingdon, Oxon ; New York, NY : Routledge, [2016] | Includes bibliographical references and index.
Identifiers: LCCN 2015050217| ISBN 9781138677944 (hbk. : alk. paper) | ISBN 9781138826571 (pbk. : alk. paper) | ISBN 9781315739236 (ebook : alk. paper)
Subjects: LCSH: Civil engineering contracts--Great Britain--Miscellanea. | Construction contracts--Great Britain--Miscellanea. | NEC Contracts.
Classification: LCC KD1651 .H8317 2017 | DDC 343.4107/869--dc23
LC record available at http://lccn.loc.gov/2015050217

ISBN: 978-1-138-67794-4 (hbk)
ISBN: 978-1-138-82657-1 (pbk)
ISBN: 978-1-315-73923-6 (ebk)

Typeset in Goudy
by HWA Text and Data Management, London

Contents

Figures and tables

Figures

Tables

Preface

I have been involved with the NEC family of contracts since 1995 and in that time I have advised on numerous projects using the contracts, and carried out over 1,500 NEC based training courses. I was Secretary of the NEC Users' Group from 1996 to 2006, providing support (including seminars and workshops) to users of the contracts, during which time I also ran the Users' Group Helpline answering queries on the NEC contract from members of the Group.

During this involvement with NEC contracts and associated training courses, I have been asked a multitude of questions, some of which are unique, but many falling into the category of "frequently asked questions" – though, of course, all of these questions are important to the questioner and when asked require a constructive and practical answer.

I have always felt that it would be useful to NEC practitioners if I could gather up a collection of these questions and also include the answers, but the problem is how many questions should be considered, and on that point I felt 100 questions and answers would be a reasonable number, though I can think of many more!

In dealing with the questions and answers I have primarily used the NEC3 Engineering and Construction Contract (ECC) as the principal reference document. Many of the answers also related to other members of the NEC3 family; therefore, I have (where relevant) specifically referred to the following:

- NEC3 Engineering and Construction Short Contract (ECSC)
- NEC3 Professional Services Contract (PSC)
- NEC3 Term Service Contract (TSC) and Term Service Short Contract (TSSC)
- NEC3 Supply Contract (SC) and Supply Short Contract (SSC).

Occasionally, I have referred to each of the NEC3 contracts using the above abbreviations, although I tend not to use too many abbreviations, preferring to refer to the contracts by their correct title.

References to clause numbers throughout this book relate to the April 2013 NEC3 reprints, though the clause numbering is not dissimilar to the original NEC contracts and subsequent NEC3 June 2005 versions.

Whilst a number of practitioners still use the NEC 1st and 2nd editions of the NEC contracts, this book is primarily aimed at giving guidance to NEC3 users, though the structure and contents of the contracts are, again, very similar and to that end much of the advice given in this book is of use to all NEC users.

The book is intended to be of benefit to professionals who are actually using the contract, but also to students who need some awareness of the contract as part of their studies.

Readers may note the absence of case law within this text. As with my previous three NEC books, this is a deliberate policy on my part for three reasons.

First, I am not a lawyer, my background being in senior commercial positions within major building contractors and currently in senior management with a major project management company in Qatar, so I felt, and readers may concur, that I was unqualified to quote and to attempt a detailed commentary on any case law.

Second, there has actually been very little case law on the NEC contracts since they were first launched. This is good news in the sense that NEC has largely avoided resorting to the "tribunal" within the contracts, but lawyers will often refer to and relay on case law and precedent, and in that case I am afraid, but at the same time pleased, that they will find very little with reference to the NEC contracts.

Third, and probably the most important, as a contracts consultant with significant overseas experience of all contracts including the NEC, it was always my intention that the book should attract an international readership. NEC was always conceived as an international contract, so including UK case law would probably limit it to a UK readership.

Any references within the book to "Acts" refers to Acts in force within the UK.

Kelvin Hughes
December 2015

Acknowledgements

I would like to extend my sincere thanks and gratitude to my wife Lesley, who as always gives me the love, the time, the inspiration and the support to fulfil my life's ambitions!

This book is especially dedicated to my young granddaughter Emily Mair Hughes ("Little Miss Sunshine") who seems to have been born with a smile on her face, bringing such joy to her parents Andrew and Louise and to her whole family. As she grows up, may she continue to bring that joy to others, and to experience the warmth and support of our wonderful family, who have always supported me.

Introduction to the NEC3 Contracts

Question 0.1 We are compiling tender documents for a project we have valued at approximately £500,000, and are wondering whether we should consider using the NEC3 Engineering and Construction Contract or the Engineering and Construction Short Contract. What is the difference between the two, and is there any project for which the NEC3 contracts would not be applicable?

Many practitioners will assume that for larger projects over a certain value, the Engineering and Construction Contract would be used, and for a lower value project the Engineering and Construction Short Contract would be used, but that assumption would be incorrect, as one would also need to consider the complexity of the work, the risk associated with that work, and the need for various provisions that would be in one contract but not the other.

As the first page of the NEC3 Engineering and Construction Short Contract states:

> The NEC3 Engineering and Construction Short Contract is an alternative to Engineering and Construction Contract and is for use with contracts which do not require sophisticated management techniques, comprise straightforward work and impose only low risks on both the Employer and the Contractor.

In some ways the choice may be linked to project value in that smaller projects would be less complex, the risks would be low, and there would be a need for fewer provisions, i.e. a shorter contract.

In order to consider this question in more detail in terms of the use of the Engineering and Construction Contract or the Engineering and

Construction Short Contract, and the applicability generally of NEC3 contracts, let us first briefly review the two contracts.

Engineering and Construction Contract

The original objectives of the NEC contracts, and more specifically the Engineering and Construction Contract, were to make improvements over other contracts under three headings:

(i) Flexibility

The contract would be able to be used as follows:

- For any engineering and/or construction work containing any or all of the traditional disciplines such as civil engineering, building, electrical and mechanical work, and also process engineering.
- Previously, contracts had been written for use by specific sectors of the industry, e.g. ICE civil engineering contracts, JCT building contracts, IChemE process contracts.
- Whether the Contractor has full, some or no design responsibility.
- Previously, most contracts had provided for portions of the work to be designed by the Contractor, but with separate design and build versions if the Contractor was to design all or most of the works.
- To provide all the normal current options for types of contract such as lump sum, remeasurement, cost reimbursable, target and management contracts.
- Previously, contracts were written, primarily as either lump sum or remeasurement contracts, so there was no choice of procurement method when using a specific standard form.
- To allocate risks to suit each particular project.
- Previously, contracts were written with risks allocated by the contract drafters, and one had to be expert in contract drafting to amend the conditions to suit each specific project on which it was used.
- Anywhere in the world.
- Previously, contracts included country specific procedures and legislation and therefore either could not be used or had to be amended to be used in other countries. To that end FIDIC contracts have been seen as the only forms of contract which could be used in a wide range of countries.

- To date, the NEC contracts have been used for projects as diverse as airports, sports stadiums, water treatment works, housing projects, in many parts of the works and even research projects in the Arctic!

(ii) Clarity and simplicity

- The contract is written in ordinary language and in the present tense.
- As far as possible, NEC only uses words which are in common use so that it is easily understood, particularly where the user's first language is not English.
- Previously obscure words such as "whereinbeforesaid", "hereinafter" and "aforementioned" were commonplace in contracts! NEC has few sentences that contain more than 40 words and uses bullet points to subdivide longer clauses.
- The number of clauses and the amount of text are lower than in most other standard forms of contract and there is an avoidance of cross-referencing found in more traditional standard forms.
- It is arranged in a format which allows the user to gain familiarity with its contents, and required actions are defined precisely thereby reducing the likelihood of disputes.
- Finally, subjective words such as "fair" and "reasonable" have been used as little as possible as they can lead to ambiguity, so more objective words and statements are used.

Some critics of NEC have commented that the "simple language" is actually a disadvantage as certain clauses may lack definition and there are certain recognised words that are commonly used in contracts. There is very little case law in existence with NEC contracts, and whilst adjudications are confidential and unreported, anecdotal evidence suggests that there do not appear to be any more adjudications with NEC contracts than any other, which would tend to suggest that the criticism may be unfounded.

(iii) Stimulus to good management

This is perhaps the most important objective of the Engineering and Construction Contract in that every procedure has been designed so that its implementation should contribute to, rather than detract from, the effective management of the work. In order to be effective in this respect, contracts should motivate the parties to proactively want to manage the outcome of the contract, not just to react to situations. NEC intends the parties to be proactive and not reactive. It also requires the parties and those that represent them to have the necessary experience (sadly often lacking) and to be properly trained so that they understand how the NEC works.

The philosophy is founded on two principles:

- "Foresight applied collaboratively mitigates problems and shrinks risk."
- "Clear division of function and responsibility helps accountability and motivates people to play their part."

Examples of foresight within the Engineering and Construction Contract are the early warning and compensation event procedures, the early warning provision requiring the Project Manager and the Contractor each to notify the other upon becoming aware of any matter which could have an impact on price, time or quality.

A view held by many Project Managers is that an early warning is something the Contractor would give, and is an early notice of a "claim". This is an erroneous view as, first, early warnings should be given by either the Project Manager or the Contractor, whoever becomes aware of it first, the process being designed to allow the Project Manager and Contractor to share knowledge of a potential issue before it becomes a problem, and, secondly, early warnings should be notified regardless of whose fault the problem is, it is about raising and resolving the problem, not compensating the affected party.

The compensation event procedure requires the Contractor to submit within three weeks a quotation showing the time and cost effect of the event. The Project Manager then responds to the quotation within two weeks, enabling the matter to be properly resolved close to the time of the event rather than many months or even years later.

The programme is also an important management document with the contract clearly prescribing what the Contractor must include within his programme and requiring the Project Manager to "buy into it" by formally accepting (or not accepting) the programme. The programme is then defined as the "Accepted Programme".

In total, the Engineering and Construction Contract is designed to provide a modern method for Employers, Contractors, Project Managers, Designers and others to work collaboratively and to achieve their objectives more consistently than has been possible using other

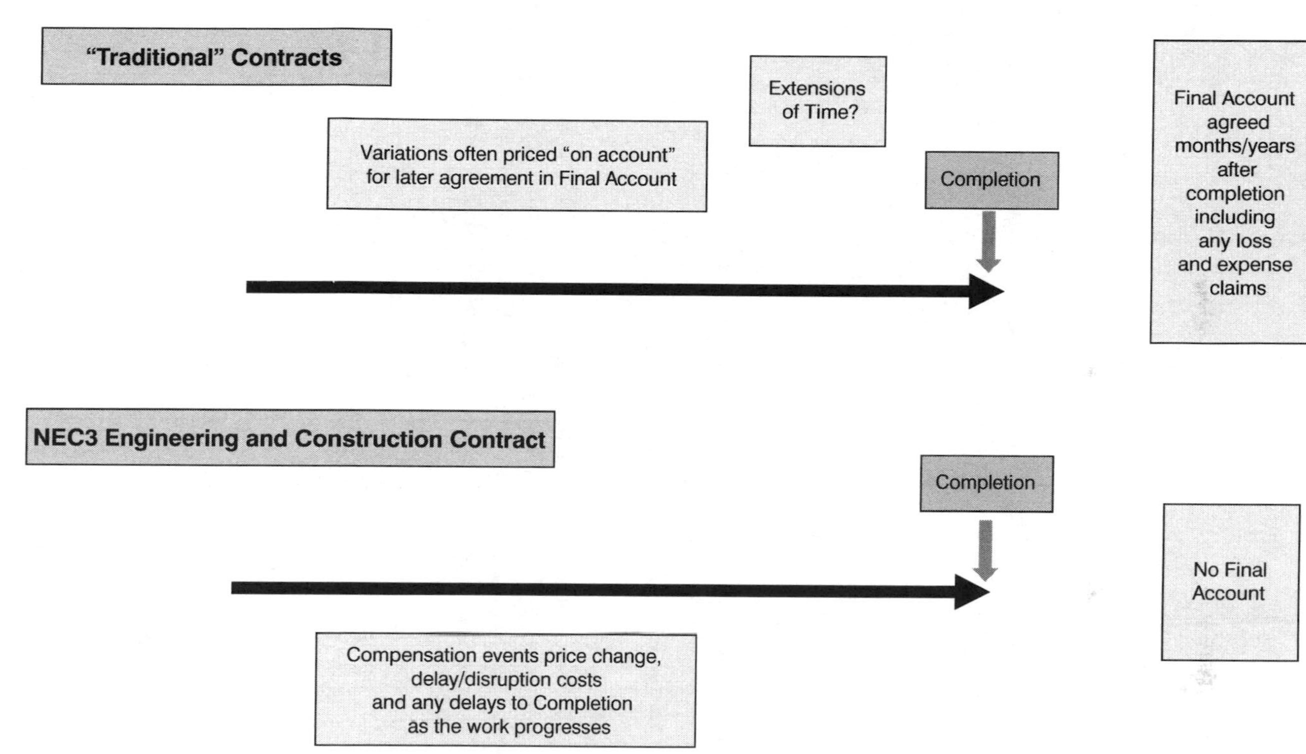

Figure 0.1 Change Management

traditional forms of contract. People will be motivated to play their part in collaborative management if it is in their commercial and professional interest to do so.

Uncertainty about what is to be done and the inherent risks can often lead to disputes and confrontation, but the Engineering and Construction Contract clearly allocates risks and the collaborative approach will reduce those risks for all the parties so that uncertainty will not arise:

- Flexibility of use – the Engineering and Construction Contract is not sector specific in terms of it being a building, civil engineering, mechanical engineering, process contract, etc.; as with other contracts, it can be used for any form of engineering or construction. This is particularly useful where a major project such as an airport or a sports stadium can be a combination of building, civil engineering and major mechanical and electrical elements.
- Flexibility of procurement – the Engineering and Construction Contract's Main and Secondary Options, together with flexibility in terms of Contractor design, allow it to be used for any procurement method whether the Contractor is to design all, none or part of the works. Again, other contracts do not offer this flexibility.
- Early warning – the Engineering and Construction Contract contains express provisions requiring the Contractor and the Project Manager to notify and, if required, call a "risk reduction" meeting, when either becomes aware of any matter which could affect price, time or quality. Few contracts have this express requirement.
- Programme – there is a clear and objective requirement for a detailed programme with method statements and regular updates which provides an essential tool for the parties to manage the project and to notify and manage the effect of any changes, problems, delays, etc. Whilst other contracts contain programme requirements, they do not deal with them in the same detail, and one could imply that they probably do not properly recognise the importance of a programme.
- Compensation events – this procedure is unique to NEC and requires the Contractor to price the time and "Defined Cost" effect of a change within three weeks and for the Project Manager to respond within two weeks. There is therefore a "rolling" Final

Account with early settlement and no later "end of job" claims for delay and/or disruption. It is also more beneficial for the Contractor in terms of his cash flow as the Contractor is paid agreed sums rather than reduced "on account" payments, which are subject to later agreement and payment.

- Disputes – the contract encourages better relationships and there is far less tendency for disputes because of its provisions. If a dispute should arise there are clear procedures as to how to deal with it, i.e. adjudication, tribunal.

Arrangement of the Engineering and Construction Contract

The Engineering and Construction Contract includes the following sections:

Core clauses

1 General
2 The Contractor's main responsibilities
3 Time
4 Testing and defects
5 Payment
6 Compensation events
7 Title
8 Risks and insurance
9 Termination

Main Option clauses

Option A	Priced contract with activity schedule
Option B	Priced contract with bill of quantities
Option C	Target contract with activity schedule
Option D	Target contract with bill of quantities
Option E	Cost reimbursable contract
Option F	Management contract

- Options A and B are priced contracts in which the risks of being able to carry out the work at the agreed prices are largely borne by the Contractor.
- Options C and D are target contracts in which the Employer and Contractor share the financial risks in an agreed proportion.

- Options E and F are two types of cost reimbursable contract in which the financial risks of being able to carry out the work are largely borne by the Employer.

Dispute resolution

Option W1	Dispute resolution procedure (used unless the Housing Grants, Construction and Regeneration Act 1996* applies).
Option W2	Dispute resolution procedure (used in the United Kingdom when the Housing Grants, Construction and Regeneration Act 1996* applies).

Secondary Option clauses

Option X1	Price adjustment for inflation
Option X2	Changes in the law
Option X3	Multiple currencies
Option X4	Parent company guarantee
Option X5	Sectional Completion
Option X6	Bonus for early Completion
Option X7	Delay damages
Option X12	Partnering
Option X13	Performance bond
Option X14	Advanced payment to the Contractor
Option X15	Limitation of the Contractor's liability for his design to reasonable skill and care
Option X16	Retention
Option X17	Low performance damages
Option X18	Limitation of liability
Option X20	Key Performance Indicators
Option Y(UK)1	Project Bank Account
Option Y(UK)2	The Housing Grants, Construction and Regeneration Act 1996
Option Y(UK)3	The Contracts (Rights of Third Parties) Act 1999
Option Z	Additional conditions of contract

Note: Options X8 to X11 and X19 are not used.

* *See amendment regarding the Local Democracy, Economic Development and Construction Act 2009.*

Note: The Local Democracy, Economic Development and Construction Act 2009 has a direct bearing on Option Y(UK)2.

The new Act, specifically Part 8, amends Part II of the Housing Grants, Construction and Regeneration Act 1996, and came into force in England and Wales on 1 October 2011 and in Scotland on 1 November 2011.

It is important to note that as the new Act amends the Housing Grants, Construction and Regeneration Act 1996, you have to take into account both Acts to fully understand how it applies to your contract.

Briefly, the differences between the Housing Grants, Construction and Regeneration Act 1996 and the Local Democracy, Economic Development and Construction Act 2009 include the following.

- A notice is to be issued by the payer to the payee within five days of the payment due date, or by the payee within not later than five days after the payment due date, the amount stated is the notified sum. The absence of a notice means that the payee's application for payment serves as the notice of payment due and the payer will be obliged to pay that amount.
- A notice can be issued at a prescribed period before the final date for payment which reduces the notified sum.
- The Contractor has the right to suspend all or part of the works, and can claim for reasonable costs in respect of costs and expenses incurred as a result of this suspension as a compensation event.
- Terms in contracts such as "the fees and expenses of the Adjudicator as well as the reasonable expenses of the other party shall be the responsibility of the party making the reference to the Adjudicator" will be prohibited. Much has been written about the effectiveness of such clauses, and whether they comply with the previous Act, so the new Act should provide clarity for the future.
- The Adjudicator is permitted to correct his decision so as to remove a clerical or typographical error arising by accident or omission. Previously he could not make this correction.

Note: The publishers have issued a brief amendment to the Engineering and Construction Contract to align it with the Local Democracy, Economic Development and Construction Act 2009.

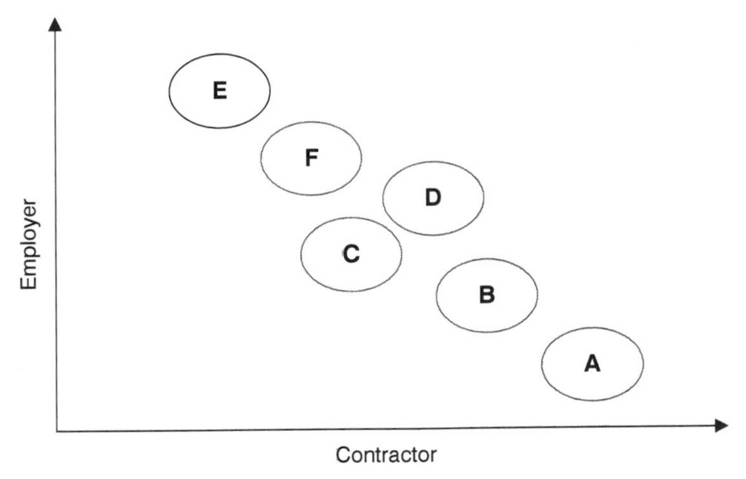

Figure 0.2 Financial Risk of Main Options

The Main Options

The six Main Options (Option A to Option F) enable Employers to select a procurement strategy and payment mechanism most appropriate to the project and the various risks involved. Essentially, the main options differ in the way the Contractor is paid (see Figure 0.2).

Whilst many traditional contracts are based on bills of quantities, there has been a movement away from the use of traditional bills and towards payment arrangements such as milestone payments and activity schedules, with payment based on progress achieved, rather than quantity of work done.

There is also an increasing use of target cost contracts which has been encouraged by the increasing use of partnering arrangements, the better sharing of risk and also the continued growth of NEC contracts which provide for target options. To that end, it is perhaps not surprising that in a survey carried out by the RICS (Contracts in Use Survey), Options A and C were found to be the most regularly used NEC3 Engineering and Construction main options.

Once the procurement strategy has been decided, the Main and Secondary Options can be selected to suit that strategy.

Engineering and Construction Short Contract

As already stated, many users of the NEC3 contracts believe that the Engineering and Construction Contract is used for larger projects and the Engineering and Construction Short Contract for smaller projects.

The majority of building and engineering contracts carried out are of relatively low risk and complexity, and often but not necessarily, low value. Hence the work involved in preparing a full Engineering and Construction Contract cannot be justified in many cases.

Also, the detailed procedures and management systems in the Engineering and Construction Contract may not be necessary in many contracts. Thus, there is a definite need for a shorter contract which is simple, easy to use and more suited to the low risk and complexity, and small value type of contract.

The Latham Report Constructing the Team emphasised the need for a shorter contract, paragraph 5.20 of the report stating that "provision should be made for a simpler and shorter minor works document".

Unlike other forms of contract, there is no limiting or recommended financial value for contracts under the Engineering and Construction Short Contract, because the above criteria are not in general related to a particular sum of money. It is possible to have large value, low risk and simple work carried out under the Engineering and Construction Short Contract. On the other hand it may not be appropriate to carry out complex high risk work under the Engineering and Construction Short Contract even though its monetary value is small. Several Employers have used the Engineering and Construction Short Contract very successfully for work in excess of £2m in value.

Intended users of the Engineering and Construction Short Contract range from major organisations carrying out a large number of straightforward, low risk projects, to domestic householders seeking a simple, user friendly contract to enable them to appoint a builder for a home extension. It has to be said that the great majority of users of the contract fall into the former category.

Since the Engineering and Construction Short Contract has been drafted for use on straightforward and low risk work, many matters covered by the Engineering and Construction Contract have been omitted. Whilst the intention has been to cater for anything which might go wrong on the simpler project, it is always possible that something may occur which is not covered in the Engineering and Construction Short Contract. Such matters will then have to be dealt with by the common law of contract. A comparison with the Engineering and Construction Contract is given below.

Differences between the Engineering and Construction Contract and the Engineering and Construction Short Contract

Detailed comparison with the clauses of the Engineering and Construction Contract is outside the scope of this book. However there are some important differences of a general nature.

For example, the Project Manager and Supervisor are absent, but the Employer has powers to delegate as necessary to suit his organisation and his particular purposes.

Also there are no Main Options; instead the flexibility and allocation of risk is covered in the Price List.

Neither are there any Secondary Options; the reason for this is that these would probably not be used in the type of contract for which the Engineering and Construction Short Contract is designed. However, retention and delay damages, both Secondary Options in the Engineering and Construction Contract, have been included in the clauses of the Engineering and Construction Short Contract.

The number of definitions has been reduced, and since there are no Main Options, the Prices and the Price for Work Done to Date have standard definitions with the flexibility being inherent in the use of the Price List.

Some clauses which are unlikely to be used have been omitted but if required can be incorporated by appropriate drafting of the Works Information or the use of additional conditions of contract. For instance, no mention is made of health and safety requirements, but any specific requirements can be included in the Works Information.

There is a requirement that the Employer is to accept the Contractor's design before he can start work on the work designed although detailed procedures for acceptance have not been included. Similarly there is no procedure for acceptance of Subcontractors, but responsibility of the Contractor for the performance of Subcontractors is clearly stated.

In the Engineering and Construction Contract there is a prescribed list of items which must be included in the Contractor's programme. All of these may not be necessary for a project using the Engineering and Construction Short Contract, and therefore any detailed requirements for the programme within the Engineering and Construction Short Contract are to be stated in the Works Information.

The payment procedure in the Engineering and Construction Short Contract is different from that in the Engineering and Construction Contract. In the Engineering and Construction Short Contract the Contractor is required to submit an application for payment after which the Employer is obliged to pay after making any corrections he considers necessary. There is no payment certificate as in the Engineering and Construction Contract.

Also the definition of Contractor's Cost is much simplified. There is no Schedule of Cost Components; instead cost is defined in terms of payments made by the Contractor for four different items:

- People employed by the Contractor
- Plant and Materials
- Work subcontracted by the Contractor, and
- Equipment.

The list of compensation events has been reduced from 19 No in the Engineering and Construction Contract to 14 No in the Engineering and Construction Short Contract. The weather event has been simplified in terms of time lost.

Assessment of compensation events is similar to that of the Engineering and Construction Contract except that effectively the rates in the Price List are used where possible.

The section on title is much abbreviated and the title to surplus materials on the site is now with the Contractor rather than the Employer.

Section 8 on insurance etc. has been completely restructured to include the equivalent of force majeure events.

Reasons which give rights to termination in Section 9 have been reduced and some of the detailed procedures on termination have been omitted.

Contract Strategy

The Engineering and Construction Short Contract does not give the Employer the choice of contract strategy in the form that is provided in the Engineering and Construction Contract.

However, the Price List can be used to produce a lump sum contract or a bill of quantities based contract. Target contracts and cost reimbursable contracts are not provided for as they are regarded as unsuitable and too complex for this type of work. Similarly management contracts will not be used on the type of work which the Engineering and Construction Short Contract is designed for.

When the Employer prepares the contract he must decide the strategy he wishes to employ and hence who should carry the risks of quantities, pricing, etc.

Question 0.2 We wish to engage a Contractor to carry out electrical testing within existing council housing stock over a period of three years. The work is fairly straightforward. Can we use the NEC3 Term Service Short Contract or should we use the Term Contract?

As stated previously, whether one uses the NEC3 Term Service Short Contract or the Term Contract is not dictated by the size of the project, but by the complexity of the work, the risk associated with that work, and the need for various provisions that would be in one contract but not the other.

In order to consider this question, let us briefly review the two contracts.

Term Service Contract

The Term Service Contract is designed for use in a wide variety of situations where a Contractor is appointed to carry out, as its name suggests, a defined service over a predetermined period of time, referred to in the contract as the "service period" starting at the "starting date".

When deciding whether to use the Term Service Contract one must recognise that this is not a contract to provide a project. The principle is that it is based on providing a service i.e. maintaining an existing condition for a period of time (term) to permit the Employer's continuing use of a facility.

It does not normally include the improvement of an existing condition of an asset – that would comprise a project. However, a modest amount of improving the condition of an asset – sometimes called "betterment" – may sometimes be sensibly included in a Term Service Contract.

Another consideration is that there is no equivalent of the "Site" or "Working Areas" as in the Engineering and Construction Contract; the contract refers to the "Affected Property".

The "Service Information" describes the service to be provided by the Contractor. It also includes full details of where and how it is to be provided and any constraints placed upon the Contractor.

Examples of its use include:

- maintenance of highways in a particular area
- periodic inspection and reporting on structures, e.g. bridges and tunnels
- cleaning of streets in an urban area
- refuse collection and disposal
- maintaining public parks and landscape areas
- maintaining mechanical and electrical installations in buildings
- maintenance of water courses such as rivers and canals
- the provision of security personnel for an installation, site or building.

Examples of other and more complex applications may be:

- servicing and maintaining airport and sea terminal buildings
- maintaining lifts in a group of hospitals
- the provision of data processing services by a computer systems company for a number of years
- carrying out long-term maintenance as part of facilities management contracts.

The form and structure of the Term Service Contract is similar to other NEC3 contracts so users will be immediately familiar with early warning, compensation events, etc.

The Employer is represented by a "Service Manager" who administers the contract on behalf of the Employer in the same way that the Project Manager and Supervisor do in the Engineering and Construction Contract.

Other differences between the Term Service Contract and the Engineering and Construction Contract include the following.

The Contractor's plan

The Term Service Contract replaces the Engineering and Construction Contract's programme with the Contractor's plan. The reason for this is that in a service contract much more emphasis is placed on how the Contractor proposes to provide the service throughout the service period, rather than the timing of his activities to provide the works, as in the Engineering and Construction Contract.

Affected Property

The Site and Working Areas of the Engineering and Construction Contract are replaced by the Affected Property, which is a defined term in the core clauses.

Main Option clauses

There are three Main Options, one of which must be selected. These are:

Option A	Priced contract with Price List
Option C	Target contract with Price List
Option E	Cost reimbursable contract

Each of these has a different allocation of risk between Contractor and Employer.

Option A has the greatest financial risk for the Contractor in that he is bound by the rates and prices in the Price List regardless of the actual cost to him of providing the services. The Price List is a combination of lump sum items and quantity-related items. Payments to the Contractor (defined as the Price for Services Provided to Date) consist of lump sums for those items which have been completed and amounts for the other items calculated as quantities of work completed multiplied by the rates, i.e. by admeasurement.

Option C is a cost reimbursable contract in which the Contractor is paid his Defined Cost together with the "Contractor's share".

The latter is calculated by comparing the Defined Cost at various stages in the provision of the services, with a target price (the Prices) which is calculated from the rates and prices in the Price List. The difference between the Prices and the Price for Services Provided to Date is divided into share ranges. The Contractor's share is calculated for each of the share ranges and the total of these is the total Contractor's share payable to the Contractor.

Certain "Disallowed Cost" defined in the core clauses is deducted to arrive at the Defined Cost which determines the payments to be made to the Contractor in Options C and E. The Fee is added to Defined Cost – this is intended to broadly cover the Contractor's head office overheads and profit. The target (the Prices) is adjusted to allow for compensation events as they are assessed in accordance with the contract.

Option E is a cost reimbursement contract similar to Option C but without the incentive to the Contractor in the form of the Contractor's share. It is therefore intended for use only where the risks are high and where the service to be provided by the Contractor cannot be defined with any certainty at the start of the contract. It would also be suitable where the service required is of an experimental nature. Pricing in the Price List for Option E contracts is used only for estimating and budgeting purposes.

The Price List

The Price List must be prepared for each contract. A pro-forma Price List is included in Appendix 6 of the Guidance Notes.

It will be seen from the above that the Price List consists of two kinds of entries:

- lump sum items, and
- remeasurable items.

The Price List may be compiled by either the Employer or the Contractor, with the exception of the pricing. Pricing should be completed by the Contractor or negotiated with him.

Where the Contractor is paid, say, an amount each month for the service he is providing, the description of the work covered by the sum of money is entered in the "Description" column, and the rate for each month is entered in the "Rate" column. The total number of months is entered in the "Expected Quantity" column. Thus the flexibility in payment methods inherent in the main Options of the Engineering and Construction Contract is provided in the Term Service Contract by means of the different ways in which the Price List can be used.

The Term Service Contract is intended to be used for the appointment of a supplier for a period of time to manage and provide a service.

The Term Service Short Contract is also available for more straightforward projects, alongside guidance notes, flow charts and how to guides which have been written to help users improve both the setting up and management of the Term Service Contract.

The Term Service Contract contains the core clauses, the three Main Option clauses, Secondary Option clauses and Contract Data forms.

Core clauses

1 General
2 The Contractor's main responsibilities
3 Time
4 Testing and Defects
5 Payment
6 Compensation events

Ref.	Description	Quantity	Rate	Price
Sample Price List				
	Lump Sum prices			
	Employer completes Ref. and Description columns			
	Contractor completes Price column only			
	Remeasurable prices			
	Employer enters Ref., Description and Quantity columns			
	Contractor completes Rate and Price columns			

Figure 0.3 Sample Price List

7 Use of equipment, Plant and Materials
8 Risks and insurance
9 Termination

Main Option clauses

Option A	Priced contract with Price List
Option C	Target contract with Price List
Option E	Cost reimbursable contract

Dispute resolution

Option W1
Option W2

Secondary Option clauses

Option X1	Price adjustment for inflation
Option X2	Changes in the law
Option X3	Multiple currencies
Option X4	Parent company guarantee
Option X12	Partnering
Option X13	Performance bond
Option X18	Limitation of liability
Option X19	Task Order
Option X20	Key Performance Indicators
Option Y(UK)1	Project Bank Account
Option Y(UK)2	The Housing Grants, Construction and Regeneration Act 1996
Option Y(UK)3	The Contracts (Rights of Third Parties) Act 1999
Option Z	Additional conditions of contract

Note: Options X5 to X11 and X14 to X17 are not used.

- Price List
- Contract Data
- Index

The NEC3 Term Service Short Contract should be used for the appointment of a supplier for a period of time to manage and provide a service. This contract is an alternative to the NEC3 Term Service Contract and is for use with contracts which do not require sophisticated management techniques, comprise straightforward work and impose only low risks on both the Employer and the Contractor.

Term Service Short Contract

As stated, the NEC3 Term Service Short Contract should be used for the appointment of a supplier for a period of time to manage and provide a service. This contract is an alternative to the NEC3 Term Service Contract and is for use with contracts which do not require sophisticated management techniques, comprise straightforward work and impose only low risks on both the Employer and the Contractor.

It is designed for use in a wide variety of situations; examples of its use may be simpler versions of the previous list for the Term Service Contract:

- maintenance of highways in a particular area
- periodic inspection of bridges and reporting
- cleaning of streets in an urban area
- refuse collection and disposal
- maintaining public parks and landscape areas
- maintaining heating, lighting and ventilation of buildings
- maintenance of a canal and servicing the leisure facilities it affords
- snow clearing
- the provision of security personnel for an installation, site or building.

Nature of the contract

The Term Service Short Contract is a contract by which a Contractor provides a service to an Employer for a period of time (a term) which is called the "service period". This period begins at the "starting date".

The contract has been drafted using the same principles as in other contracts in the family of NEC3 documents. It uses similar procedures and wording as much as possible.

Complete package

The Term Service Short Contract is published as a complete package including:

- pre-printed forms for:
  - Contract Data
  - Contractor's Offer
  - Employer's Acceptance

- pre-printed forms for:
  - Price List
  - Service Information
  - Task Order
- Conditions of Contract

When all the forms have been completed for a particular contract the package will comprise the complete contract document, together with the drawings and anything referred to in the Service Information.

The most important notes on the use of the document are included in grey-bordered boxes.

The Parties

The main parties referred to in the Term Service Short Contract are the:

- Employer
- Employer's Agent (if appointed)
- Contractor
- Adjudicator

The Employer and the Contractor are the parties to the contract.

The Employer/Employer's Representative

Unlike the Engineering and Construction Contract, the Term Service Short Contract does not include the roles of Project Manager and Supervisor, all actions being between the Employer (or his delegated representative) and the Contractor.

If the Employer appoints a representative it can be from either from his own staff, or from an outside body. His role is to manage the contract for the Employer to achieve the Employer's objectives for the completed project. He is normally appointed at the feasibility stage of the project, his duties then including advising on design, procurement, cost planning and programme matters.

If the Employer has set limits upon his level of authority, for instance agreeing the value of compensation events, he must ensure that there is an efficient and speedy authorisation procedure to allow him to exceed these limits.

The Adjudicator

The Adjudicator is appointed jointly by the Employer and the Contractor for the contract. The name of the Adjudicator is inserted into the Contract Data. If the Contractor does not agree with the choice a suitable alternative should be appointed before the Contract Date.

The Adjudicator becomes involved only when either contracting party refers a dispute to him. As an independent person he is required to give a decision on the dispute within stated time limits. If either party does not accept his decision they may proceed to the tribunal (litigation or arbitration). Payment of the Adjudicator's fees is shared by the Parties.

Question 0.3 We wish to purchase large quantities of rock salt for treating roads during winter periods. Is the NEC3 Supply Short Contract sufficient or should we use the Supply Contract because of the large quantities?

The NEC3 Supply Contract (SC) is used for the procurement, supply and delivery of high value goods and associated services, which can range through items such as transformers, turbines, trains, process plant, etc., together with related services which may be required as part of the contract such as design and also specific delivery requirements.

"Procurement and supply" means that these items are obtained and delivered to a Delivery Place on a Delivery Date, but there is no fixing/installation element to the contract.

What is classed as "high value" is not based on a finite figure, it is really the decision of the Employer, but there is a need for a comprehensive contract which will cover all aspects of the procurement and supply process for those goods and services.

The Parties within the NEC3 Supply Contract are the Purchaser and the Supplier, with a Supply Manager appointed and named in the Contract to represent the Purchaser.

The structure is very similar to the other NEC3 contracts, so there is provision for early warnings, programmes submitted for acceptance, and also compensation events to manage changes.

Core clauses

1 General
2 The Supplier's main responsibilities

3 Time
4 Testing and Defects
5 Payment
6 Compensation events
7 Title
8 Risks, liabilities, indemnities and insurance
9 Termination and dispute resolution

There are no Main Options, but the following Secondary Options:

Option X1	Price adjustment for inflation
Option X2	Changes in the law
Option X3	Multiple currencies
Option X4	Parent company guarantee
Option X7	Delay damages
Option X12	Partnering
Option X13	Performance bond
Option X14	Advanced payment to the Supplier
Option X17	Low performance damages
Option X20	Key Performance Indicators
Option Y(UK)1	Project Bank Account
Option Y(UK)3	The Contracts (Rights of Third Parties) Act 1999
Option Z	Additional conditions of contract

Note: Main Options A to G and Secondary Options X5, X6, X8 to X11, X15, X16, X18 and X19 used in other NEC3 contracts are not used in this contract.

The NEC3 Supply Short Contract (not, as often quoted, the "Short Supply Contract") should be used for procurement of goods probably under a single order or on a batch order basis, and is for use with contracts which do not require sophisticated management techniques and impose only low risks on both the Purchaser and Supplier.

The principle of the complexity of the requirements and whether the risks are high or low is always the deciding factor as to whether to use an NEC3 contract or its "Short Contract" version, rather than simply the value of the contract, as is so often the case with other contract families. It is not about value, it is about what you are trying to do and what factors need to be considered in doing it and achieving those objectives.

The NEC3 Supply Short Contract is structured very similar to the other NEC3 Short Contracts, i.e.:

- Contract Data
- The Supplier's Offer
- The Purchaser's Acceptance
- Price Schedule
- Goods Information
- Conditions of Contract.

Again, as with the other Short Contracts there is shortened provision for early warnings, programmes submitted for acceptance, and also compensation events to manage changes.

Question 0.4 Do the NEC3 contracts comply with the requirements of the Housing Grants, Construction and Regeneration Act 1996 and the Local Democracy, Economic Development and Construction Act 2009?

If one just examines the Core Clauses of the NEC3 contracts, the simple answer is no, the NEC3 contracts do not comply with the requirements of the Housing Grants, Construction and Regeneration Act 1996 and Local Democracy, Economic Development and Construction Act 2009, and the simple reason is because the NEC3 contracts are for use anywhere in the world, therefore they do not include any country specific clauses.

However, Main Option Y(UK)2: The Housing Grants, Construction and Regeneration Act 1996 is applicable to UK contracts where the Housing Grants, Construction and Regeneration Act 1996 applies.

Note: Option Y(UK)2 only deals only with the payment aspects of the Act; adjudication under the Act is covered by Option W2.

Whilst this book avoids discussing NEC3 specifically in connection with UK law, it is worth mentioning the Local Democracy, Economic Development and Construction Act 2009 which has a direct bearing on Option Y(UK)2. The new Act, specifically Part 8, amends Part II of the Housing Grants, Construction and Regeneration Act 1996, and came into force in England and Wales on 1 October 2011 and in Scotland on 1 November 2011.

It is important to note that as the new Act amends the Housing Grants, Construction and Regeneration Act 1996, you have to take into account both Acts to fully understand how it applies to your contract.

The Act applies to all construction contracts, including:

- construction, alteration, repair, maintenance, etc.
- all normal building and civil engineering work, including elements such as temporary works, scaffolding, site clearance, painting and decorating
- consultants' agreements concerning construction operations
- labour only contracts
- contracts of any value.

It excludes:

- extraction of oil, gas or minerals
- supply and fix of plant in process industries, e.g. nuclear processing, power generation, water or effluent treatment
- contracts with residential occupiers
- header agreements in connection with PFI contracts
- finance agreements.

The Housing Grants, Construction and Regeneration Act 1996 only applied to contracts in writing, but the Local Democracy, Economic Development and Construction Act 2009 now also applies to oral contracts.

Briefly, the differences between the Housing Grants, Construction and Regeneration Act 1996 and the Local Democracy, Economic Development and Construction Act 2009 include the following:

- A notice is to be issued by the payer to the payee within five days of the payment due date, or by the payee within not later than five days after the payment due date, the amount stated is the notified sum. The absence of a notice means that the payee's application for payment serves as the notice of payment due and the payer will be obliged to pay that amount.
- A notice can be issued at a prescribed period before the final date for payment which reduces the notified sum.
- The Contractor has the right to suspend all or part of the works, and can claim for reasonable costs in respect of costs and expenses incurred as a result of this suspension as a compensation event.
- Terms in contracts such as "the fees and expenses of the Adjudicator as well as the reasonable expenses of the other party shall be the responsibility of the party making the reference to the Adjudicator" will be prohibited. Much has been written about the effectiveness of such clauses, and whether they comply

with the previous Act, so the new Act should provide clarity for the future.

- The Adjudicator is permitted to correct his decision so as to remove a clerical or typographical error arising by accident or omission. Previously he could not make this correction.

Note: The publishers have issued a brief amendment to the Engineering and Construction Contract to align it with the Local Democracy, Economic Development and Construction Act 2009.

Question 0.5 We wish to use the NEC3 contracts for a series of projects. Can we assume that the contracts include all the necessary ancillary documents to be incorporated into the contract, for example:

- **Performance bond**
- **Parent Company Guarantee**
- **Novation Agreement**
- **Collateral Warranty**
- **Retention Bond?**

The simple answer to this question is that the NEC3 contracts do not have the ancillary documents themselves which would need to be incorporated into an NEC3 contract.

Whilst many NEC3 practitioners may allege that this is a clear and major oversight on the part of the NEC3 drafters, one must always remember that the NEC documents when first conceived were always intended to be for use on projects worldwide and in that sense it is impossible to include every conceivable ancillary document for every contract, aligned with the law in every country.

Let us consider each of the above in turn, using the NEC3 Engineering and Construction as a reference document.

Performance bond

A performance bond is an arrangement whereby the performance of a contracting party (the Principal) is backed by a third party (the Surety), which could be a bank, insurance company or other financial institution that, should the Principal fail in his obligations under the contract (normally due to the Principal's insolvency), the Surety will pay a pre-agreed sum of money to the other contracting party (the Beneficiary).

Option X13 provides for the Contractor to give a performance bond to the Employer, provided by a bank or insurer. The bank or insurer which provides the performance bond must be accepted by the Project Manager.

If a performance bond is required from the Contractor it should be provided by the Contract Date. The amount of the bond, usually 10 per cent of the contract value, must be stated in Contract Data Part 1 and the form of the bond must be in the form stated in the Works Information. Normally, the value of the performance bond does not reduce, but they should have an expiry date, which could be completion of the project, the Defects Date or may even include the six, ten or twelve year limitation period following completion of the works to cover any liability for potential latent defects, the expiry date being defined within the Works Information.

Parent Company Guarantee

This form of guarantee is given by a parent company (or holding company) to guarantee the proper performance of a contract by one of its subsidiaries (the Contractor) who, whilst having limited financial resources himself, may be owned by a larger financially sound parent company.

In most cases, it is the ultimate parent company that provides the guarantee, but sometimes, particularly when the ultimate parent company is in another country, the parent company may just be a company further up the chain within the group, perhaps the national parent who has sufficient assets to provide the required guarantee.

Option X4 provides that if a parent company owns the Contractor, the Contractor gives the Employer a guarantee by the parent company in the form set out in the Works Information. If a parent company guarantee is required, it may either be provided by the Contract Date or within four weeks of the Contract Date.

Parent company guarantees are normally used as an alternative to a performance bond (Option X13).

Such a guarantee is cheaper than a performance bond, as the Contractor will normally just charge an administration fee rather than the case of performance bond, where the Contractor is actually paying a premium for an insurance policy, but it may give less certainty of redress because it is not supplied by an independent third party so it is dependent on the survival, and the ability to pay, of the parent company.

If the Contractor has a parent company, they must procure a Parent Company Guarantee in the specimen form provided in the Works Information, for the benefit of the Employer.

A parent company is defined as an entity:

1 of which the Contractor (or any member if the Contractor is an Association of Persons) is a branch, subsidiary or other similar entity, or
2 which directly or indirectly exercises management control over the Contractor (or any member if the Contractor is an association of persons).

If the Contractor is an Association of Persons, each member of that Association must procure a Parent Company Guarantee for the benefit of the Employer.

Advance Payment Guarantee

An Advance Payment Guarantee is provided by a bank (the "Guarantor"), and guarantees that the Contractor will repay any amounts due to the Employer in the event that the Contractor fails to repay an advance payment provided by the Employer. If the Contractor is in breach of its obligations, and subject to a notice from the Employer confirming that breach and making a demand, the bank will pay a predetermined sum to the Employer without the Employer having to prove entitlement to the predetermined amount.

Novation Agreement

Whilst the principle of design and build agreements is that the Employer prepares his requirements and sends them to the tendering Contractors and they prepare their proposals to match the Employer's requirements, in reality, in nearly half of design and build contracts the Employer has already appointed a design team which prepares feasibility proposals and initial design proposals before the tenders are invited. Outline planning permission and sometimes detailed permission may have also been obtained for the scheme before the Contractor is appointed.

Each Contractor then tenders on the basis that the Employer's design team will be novated or transferred to the successful tendering Contractor who will then be responsible for appointing the team and completing the design under a new agreement.

This process is often referred to as "novation", which means "replace" or "substitute", and is a mechanism where one party transfers all its obligations and benefits under a contract to a third party. The third party effectively replaces the original party as a party to that contract, so the Contractor is in the same position as if he had been the Employer from the commencement of the original contract.

Many prefer to use the term "consultant switch" where the design consultant "switches" to work for the Contractor under different terms as a more accurate definition.

This approach allows the Employer and his advisers time to develop their thoughts and requirements, consider planning consent issues, then when the design is fairly well advanced, the Designers can then be passed to the successful Design and Build Contractor.

The NEC3 contracts do not include pro forma novation agreements but, if used, it is critical that the wording of these agreements is carefully considered, as there are many badly drafted agreements in existence. Novation can be by a signed agreement or by deed.

As with most contracts, there should be consideration, which is usually assumed to be the discharge of the original contract and the original parties' contractual obligations to each other. If the consideration is unclear, or where there is none, the novation agreement should be executed as a deed.

In many cases the agreement states briefly (and very badly) that from the date of the execution of the novation agreement, the Contractor will take the place of the Employer as if he had employed the Designers from the beginning. The document states that in place of the word "Employer" one should read "Contractor".

However, the issues of design liability, inspection, guarantees and warranties may not always apply on a back-to-back basis, so one must take care to draft the agreement in sufficient detail and refer to the correct parties.

In practice, the best way is to have an agreement drafted between the Employer and the Designer covering the pre-novation period, and a totally separate agreement drafted between the Contractor and the Designer for the post-novation period.

When a contract is novated, the other (original) contracting party must be left in the same position as he was in prior to the novation being made. Essentially, a novation requires the agreement of all three parties.

As this book is intended as a guide for NEC3 practitioners worldwide, it is not intended to examine specific legal cases within the UK, as they may not apply on an international basis, but there is certainly case law available in terms of design errors in the pre-novation phase being carried through as the liability of the Contractor in the post novation stage, and also the accuracy and validity of site investigations and the Employer's and Contractor's obligations in terms of checking and taking ownership of such information.

Collateral Warranty

There is no provision for collateral warranties within the Engineering and Construction Contract, but the Professional Services Contract provides, under Option X8 for the Consultant to enter into a collateral warranty agreement.

A collateral warranty is an agreement associated with another contract, collateral meaning "additional but subordinate" or "running side by side", the collateral warranty being entered into by a party to the primary contract.

Collateral warranties are different from traditional contracts, in that the two parties to the warranty do not have any direct commercial relationship with each other. They are useful for creating ties between third parties, but the main disadvantage of collateral warranties is the expense of providing them, particularly where there are many interested third parties such as with housing and commercial properties. When one also considers that parties can change their names or business status during the life of a warranty, then even a medium-sized project can involve 30 or more warranties.

Different parties will also have different requirements for collateral warranties, therefore the precise terms of the collateral warranty agreement need to be carefully considered in each case so that they are drafted to provide the required protection.

Collateral warranties are executed as a "deed", also referred to as "under seal", which differs from a simple contract in two respects:

1. In English law the time after a breach of contract has occurred within which one party can sue another is 12 years, as opposed to 6 years for a simple contract, which is an agreement that is executed "under hand".
2. Whereas under English law consideration is needed for a contract to be effective, this is not the case with a deed. For a company to execute a deed effectively, the document must be signed by two directors or a director and the company secretary, unless the Articles of Association contain some other requirement.

The basic ingredients of a collateral warranty are usually:

1 Definitions
2 The warranty
3 Naming of the parties
4 Prohibited materials
5 Copyright
6 Insurances
7 Assignment
8 Step-in rights
9 Law and jurisdiction.

Collateral warranties are a relatively recent feature of contracts, which have come into use in the past 20 years following some notable court cases in which it was judged not possible to recover damages for negligence in relation to Defects in construction projects, as it was held to be economic loss, which is not recoverable in tort.

Consequently, such claims had to be brought as claims for breach of contract, where economic loss is recoverable. It therefore became necessary to establish a contract between the parties involved in a construction project, such as Designers or Subcontractors, and the parties such as funders, present and future purchasers, who are not parties to the primary contracts but who have an interest in the project, hence the "collateral warranty".

1 Definitions

There will be a brief description of the services with the collateral warranty. The "contract" is the contract to which the collateral warranty agreement is collateral and the date should be specified. Since the obligations of the Consultant to the Beneficiary parallel the obligations of the Consultant to the Employer under the contract for the services, the Beneficiary should always ask to see a copy of the contract. If, for example, there are restrictions on the Consultant's liability in the principal contract, that might affect the rights of the Beneficiary under the collateral warranty.

A copy of the principal contract may be attached to the collateral warranty agreement.

2 The warranty

The Consultant gives warranties to the Beneficiary that he has exercised and will continue to exercise reasonable skill and care in his obligations to the Employer under the principal contract, including compliance with any design obligations, performance requirements and correction of Defects.

The warranty should contain a statement that the Warrantor shall have no greater liability to the Beneficiary than they would have under the primary contract.

The Consultant does not have any liability to the Beneficiary if he fails to complete the services under the contract on time, as that will usually be covered by delay damages under the principal contract.

There should be a statement that the Warrantor shall have no liability under warranty in any proceedings commenced more than 12 years after the date of Completion of the services.

3 Naming of the parties

The named parties can, for example, be the Consultant and the end user, a Subconsultant and the Employer, or a Consultant and the Employer.

The party that gives the warranty is known as the Warrantor; the other party is known as the Beneficiary.

4 Prohibited materials

The Contractor may be required to give a warranty that he will not specify or use materials which are known to be prohibited or deleterious to health and safety.

In addition, the use of deleterious materials may contravene relevant legislation or regulations as well as being expressly prohibited in the contract between the Developer and the Contractor.

5 Copyright

Normally, the Beneficiary under a collateral warranty agreement will have a right to use designs and other documents prepared by the Consultant but only in connection with the project for which those design documents are prepared, copyright to these documents remaining with the Consultant.

6 Insurances

There is normally a requirement for the Warrantor to maintain insurances including Professional Indemnity (PI) insurance, where applicable, for a stated minimum level of cover and for a period of 12 years after Completion of the Services.

7 Assignment

There is normally a provision allowing the Beneficiary up to two assignments of the warranty.

8 Step-in rights

Step in rights can allow a funder to take over the Employer's role and become the Employer under the primary contract. This is particularly important where the Employer is unable to pay the Consultant. In this case, the Consultant will have to notify the Beneficiary before he terminates the contract, thus allowing the Beneficiary to "step in" and take the Employer's place, with an obligation to pay the Consultant any money that is outstanding. The Beneficiary does not have any obligation to step in, only the right to do so. As the effect of this clause is to amend the principal contract, it is important that when the collateral warranty contains such a clause, the Employer should be a party to that document.

9 Law and jurisdiction

It is important to specify the law of the contract and the warranty and the jurisdiction of the courts that will have jurisdiction in any possible dispute.

Retention Bond

On particularly large projects it may be more practical to release retention monies early with a bond in place to protect the parties in the event of defects arising.

A Retention Bond is provided by a bank, and guarantees that if contract retention sums are paid to the Contractor early and the Contractor is in breach of its obligations, and subject to a notice from the Employer confirming that breach and making a demand, the bank will pay a predetermined sum to the Employer without the Employer having to prove entitlement to the predetermined amount.

Question 0.6 We wish to appoint a Contractor to carry out some work with the Employer's design team on pre-construction period value engineering and buildability options for a proposed project, for which we may appoint him to be our eventual Contractor, or we may decide to carry out a selective tendering process to select our Contractor. Can we do this with an NEC3 contract?

Probably the easiest way to employ the Contractor for the initial pre-construction period is by appointing him as a Consultant using the NEC3 Professional Services Contract or the Professional Services Short Contract.

The Contractor is providing a service, which can be defined within the Scope, which specifies and describes the service and states any constraints on how the Contractor, who will actually construct the project, can then later be appointed direct by negotiation or by an open or selective tendering process.

Although this sounds a very simple process, procurement is rarely that simple, and obviously in suggesting this option the Employer must consider, dependent on location, the relevant procurement law which may apply and will restrict how Consultants, Contractors and others are selected and appointed particularly for public funded projects.

This aspect and the relevant law is outside the scope of this book, but for example within the European Union the European Union (EU) Procurement Directives apply to all procurement within the public sector, above minimum monetary thresholds which are reviewed on a regular basis, and subject to EU wide principles of non-discrimination, equal treatment and transparency, the purpose of the procurement rules being to open up the public procurement market and to ensure the free movement of supplies, services and works within the EU.

In addition to the EU Member States, the benefits of the EU public procurement rules also apply to a number of other countries outside Europe because of an international agreement negotiated by the World Trade Organization (WTO) titled the "Government Procurement Agreement" (GPA).

Many believe that the NEC3 Professional Services Contract is just for appointing Consultants. However, the contract itself clearly states "This contract is for the appointment of a supplier to provide professional services", and that "supplier" can be a Consultant, a Contractor or even a Supplier in the normal sense of the word of goods and materials, particularly if they are specialist such as suppliers of electrical switchgear, process plant, specialist glazing, etc.

The structure of the Professional Services Contract is very similar to other NEC3 contracts, so there is provision for choice of Main and Secondary Options, early warnings, programmes submitted for acceptance, and also compensation events to manage changes.

Core clauses

1 General
2 The Parties' main responsibilities
3 Time
4 Quality
5 Payment
6 Compensation events
7 Rights to Material
8 Indemnity, insurance and liability
9 Termination

Main Option clauses

Option A	Priced contract with Activity Schedule
Option C	Target contract
Option E	Time-based contract
Option G	Term contract

Secondary Option clauses

Option X1	Price adjustment for inflation
Option X2	Changes in the law
Option X3	Multiple currencies
Option X4	Parent company guarantee
Option X5	Sectional Completion
Option X6	Bonus for early Completion
Option X7	Delay damages
Option X8	Collateral warranty agreements
Option X9	Transfer of rights
Option X10	Employer's Agent
Option X11	Termination by the Employer

Option X12	Partnering
Option X13	Performance bond
Option X18	Limitation of liability
Option X20	Key Performance Indicators
Option Y(UK)1	Project Bank Account
Option Y(UK)2	The Housing Grants, Construction and Regeneration Act 1996
Option Y(UK)3	The Contracts (Rights of Third Parties) Act 1999
Option Z	Additional conditions of contract

Note: Options X14 to X17 and X19 are not used.

Question 0.7 We have been using the NEC3 Engineering and Construction Contract for a number of years, and have used all the Main Options except Option D (target contract with bill of quantities), and wonder, why would anyone use that?

Option D (target contract with bill of quantities) is certainly one of the lesser used of the six Main Options.

As with Option B (priced contract with bill of quantities), the Bills of Quantities are prepared by the Employer.

This option is normally used where the Employer knows what he wants, and is able to define it clearly through the Works Information and measure it within the Bills of Quantities, but there are likely to be changes in the quantities that may or may not be considered as compensation events.

The Contractor tenders a price based on the Bills of Quantities. This price, when accepted, is then referred to as the "target". The original target is referred to as the Total of the Prices at the Contract Date.

The assessment of payments is the same as for an Option C contract:

- The target price includes the Contractor's estimate of Defined Cost plus other costs, overheads and profit to be covered by his Fee.
- The Contractor tenders his Fee in terms of percentages to be applied to Defined Cost.
- During the course of the contract, the Contractor is paid Defined Cost plus the Fee.
- The target is adjusted for compensation events and also for inflation (if Option X1 is used).

- On Completion, the Contractor is paid (or pays) his share of the difference between the final total of the Prices and the final Payment for Work Done to Date according to a formula stated in the Contract Data. If the final Payment for Work Done to Date is greater than the final total of the Prices, the Contractor pays his share of the difference.
- As with Option B the target is generated based on the Bill of Quantities, though changes which are compensation events are inserted into the Bill of Quantities as lump sums rather than on a remeasurement basis.
- The target is adjusted for compensation events and also for inflation (if Option X1 is used).

The concern with Option D amongst Employers is that as purpose of the bills of quantities is essentially just as a tender document to create the target, are they spending time and money producing a bill of quantities yet, once the Contractor is appointed and the target is established, the bill of quantities plays only a secondary part as it is not used for payments?

Question 0.8 We would like to use the Engineering and Construction Contract to carry out a project using Construction Management as a procurement method. We note that Option F is a Management Contract, but there is no Construction Management Option. How can we do this?

The principle with the Construction Management procurement method is that the Employer has direct contracts with the various "package contractors" otherwise known in other contracts as "Works Contractors" or "Trade Contractors" and he has a party such as an Employer's Agent or Representative to act on his behalf in dealing with programme, instructions, payments, changes, etc.

It is correct that there is no Construction Management Option, each of the Engineering and Construction Contract Main Options assumes that the parties to the contract are a single Employer and a single Contractor to build a project.

However, it is possible to create a Construction Management arrangement by using the NEC3 Engineering and Construction Contract as the contract between the Employer and each Contractor, the collective assembly of all the Contractors being appointed to build the single project.

The Employer's representative can then be employed as a Consultant under the NEC3 Professional Services Contract and be named as the Project Manager within each contract between the Employer and the Contractor.

Question 0.9 We wish to use the NEC3 contracts for a number of international building and civil engineering projects in various locations around the world. Is that possible?

The NEC contracts were always conceived as being for use on any type of project (buildings, civil engineering, process engineering, etc.) in any location throughout the world (see previous questions), hence the use of non-country specific drafting throughout the contracts which is extended into all the members of the current NEC3 family.

An example of country specific drafting is the JCT contracts and their specific relevance to UK-based projects, and the law of England and Wales (they have to be amended for use outside these countries) including within their provisions for UK health and safety (CDM), fair contract (Housing Grants, Construction and Regeneration Act 1996 and the Local Democracy, Economic Development and Construction Act 2009) and other legislation.

With the NEC3 contracts, Employers in any country can include Secondary Options to provide for any country specific legislation.

So the answer is, yes, the NEC3 contracts can be used for a wide range of projects in any location throughout the world and, in fact, have been used for projects as diverse as airports, sports stadiums, power stations, water treatment works and housing projects in many parts of the works and even research projects in the Arctic!

Question 0.10 What is the role of the Project Manager on an NEC3 Engineering and Construction Contract and does he have an obligation to act impartially between the parties?

The Project Manager is appointed by the Employer and manages the contract on his behalf.

Other forms of contract tend to name an Engineer or Architect who acts for the Employer and, in addition to having design responsibilities, also administers the contract. The Engineering and Construction Contract separates the role of design from that of managing the project and administering the contract.

The Project Manager is named in Contract Data Part 1 and may be appointed from the Employer's own staff or may be an external consultant. The Project Manager is a named individual, not a company.

It is vital that the Employer appoints a Project Manager who has the necessary knowledge, skills and experience to carry out the role, which includes acceptance of designs and programmes, certifying payments and dealing with compensation events.

With regard to impartiality, the Project Manager's role is to act on behalf of the Employer in managing the project and the contract for him.

In that sense he is not truly impartial, but for example when called to accept or not accept something such as the Contractor's design, his proposed Subcontractor or his programme, the Project Manager either accepts or gives his reasons for not accepting, so he has a duty to comply with the contract and to act fairly and "in a spirit of mutual trust and co-operation", but in doing so to act on behalf of the Employer.

If the Project Manager is seen not to be complying with the contract, and the Contractor is dissatisfied, he obviously has the right to refer any matter to adjudication.

Disciplines which have carried out the Project Manager role to date include Engineers from a civil engineering, structural or process background, Architects, Building Surveyors and Quantity Surveyors, in addition to Project Managers themselves.

It is also vital that the Employer gives the Project Manager full authority to act for him, particularly when considering the timescales imposed by the contract. If the Project Manager has to seek approvals and consents from the Employer, then he must do so, and the appropriate authority must be given, in compliance with the contractual timescales.

If the approval process is likely to be lengthy, it is critical that both the Employer and his Project Manager set up an accelerated process to comply with the contract or that the timescales in the contract are amended. The former is the vastly preferred method in order that NEC3 will work to its full effect. Although the Project Manager manages the contract at the post-contract stage, he is usually appointed pre-contract to deal with matters such as feasibility issues, advising on design, procurement, cost planning tendering and programme matters.

Note that, unlike many other contracts, the Engineering and Construction Contract does not name the Quantity Surveyor, so this would need to be considered in dealing with financial matters pre- and post-contract.

Either the Project Manager may carry out the role of the Quantity Surveyor himself or he may delegate to another.

The role of a Project Manager within the Engineering and Construction Contract is different from the usual concept of a Project Manager and it is important that this difference is recognised, as it is a traditional area of misunderstanding amongst users of the contract.

There should only be one Project Manager for the project, though the Project Manager can delegate responsibilities to others after notifying the Contractor, and may subsequently cancel that delegation (Clause 14.2).

As with any other notification under the contract, if the Project Manager (or the Supervisor) wishes to delegate, the notice to the Contractor must be in a form which can be read, copied and recorded, i.e. by letter or email, not by verbal communication such as a telephone call. The delegation may be due to the Project Manager being absent for a period due to holidays or illness, or because he wishes to appoint someone to assist him in his duties.

Although the contract is not specific, the notice should identify who the Project Manager is delegating to, what their authority is, and how long the delegation will last, so the Contractor is in no doubt as to who has authority under the contract. It is important to recognise that in delegating the Project Manager is sharing an authority with the delegate who will then represent him; he is not passing responsibility on to them. The authority is shared, but ultimately responsibility will remain with the Project Manager as the principal.

It is also important to note that one can only delegate outwards from the principal, i.e. no one can assume authority possibly because, within an organisation, they are senior to the Project Manager. If the Employer wishes to replace the Project Manager or the Supervisor, he must first notify the Contractor of the name of the replacement before doing so.

Question 0.11 What construction discipline is best suited to being the Supervisor on an NEC3 Engineering and Construction Contract, and how much "supervising" does the Supervisor actually have to do?

In order to properly consider this question, we need first to consider the role and responsibilities of the Supervisor.

The Supervisor is appointed by the Employer, his role being to manage issues regarding testing and defects on the Employer's behalf.

The Supervisor also has another role to play in marking Plant and Materials which are outside the Working Areas; the Contractor prepares them for marking as the Works Information requires, and the Supervisor then marks them as for this contract, title, then passing to the Employer.

As with the Project Manager, the Supervisor is named in Contract Data Part 1 and may be appointed from the Employer's own staff or may be an external consultant, and again, as with the Project Manager, the Supervisor is a named individual, not a company.

It is possible that, provided he has sufficient expertise and time available, the Project Manager and the Supervisor could be the same person.

It is also not unusual for the Employer's Designer and the Supervisor to be the same person as he is well qualified, having done the design, to inspect the works to make sure the Contractor has complied with it.

The title Supervisor is not normally found in other contracts, many seeing the role as similar to that of a Clerk of Works or Resident Engineer, and in many ways the role is similar, but it is important to recognise that those roles are usually delegated by others within their respective contracts, therefore giving the Clerk of Works or Resident Engineer little or no ultimate responsibility.

The normal role of a Supervisor is to take charge of and direct people or activities, though in effect he does not actually supervise in that sense as the Contractor, for example Clause 43.1 requires that the Contractor "corrects a Defect whether or not the Supervisor notifies him of it" clearly requiring the Contractor to be proactive in dealing with Defects and not rely on the Supervisor to tell him what is defective and what to do about it, which would be the traditional role of a supervisor in a company.

Disciplines which have carried out the Supervisor role to date depend on the type of project, but they include Engineers from a civil engineering, structural or process background for highways and infrastructure projects, Architects and other design disciplines for building projects, Building Surveyors for building refurbishment projects, and Clerks of Works. It really depends on the type of work what discipline the Supervisor comes from.

It is important to recognise that the Supervisor does not act on behalf of the Project Manager; he represents the Employer and has his own responsibilities and obligations within the contract, which may be summarised as follows:

- The Supervisor carries out and/or witnesses tests being carried out by the Contractor or some other party.
- The Supervisor may instruct the Contractor to search for a Defect. This is common to other contracts where it is referred to as "opening up" or "uncovering" – the principle being if no Defect is found, the matter is dealt with as a compensation event.

- The Supervisor notifies the Contractor of each Defect as soon as he finds it.
- The Supervisor issues the Defects Certificate at the later of the defects date and the end of the last defect correction period. This is a very important action, as once the Defects Certificate has been issued, the Contractor is no longer required to provide the insurances under the contract.
- Also, dependent on which Secondary Options have been selected, any retention (Clause X16) is released back to the Contractor; if a Defect included in the Defects Certificate shows low performance, the Contractor pays low performance damages (Clause X17), and the Contractor no longer has to report performance against KPIs (Clause X20).
- The Defects Certificate is used on the later of the defects date and the last Defect correction period, not when all defects have been corrected. In this case, the Defects Certificate may show Defects which the Contractor has not corrected.
- The Supervisor marks Plant and Materials as for the contract if the contract identifies them for payment. Marking Plant and Materials can involve making a physical mark on them to denote that he has seen them but can include compiling an inventory, photographic records, etc.
- The Contractor is required to prepare the Plant and Materials for marking as required by the Works Information, which can include setting them aside from other stock, protection, insurances and any vesting requirements. Once they have been marked, any title the Contractor has to them passes to the Employer.
- Again, as with the Project Manager, there should only be one Supervisor, who may then delegate duties and authorities to others. On a large project it is common for various personnel to be responsible for checking mechanical, electrical, structural, finishings, landscape elements, etc., but again there should be one Supervisor who delegates duties to others in respect of these elements. Each then reports back to the single Supervisor.

Question 0.12 Under Clause 14.2 of the NEC3 Engineering and Construction Contract, the Project Manager or Supervisor may delegate any actions. What does this mean? Is there anything that the Project Manager or Supervisor cannot or must not delegate?

There should only be one Project Manager and one Supervisor for the project, though they can both delegate responsibilities to others after notifying the Contractor, and may subsequently cancel that delegation (Clause 14.2), any reference to an action of either the Project Manager or the Supervisor including an action by his delegate.

As with any other notification under the contract, if the Project Manager (or the Supervisor) wishes to delegate under Clause 14.2, the notice to the Contractor must be in a form which can be "read, copied and recorded", i.e. by letter or email, not by verbal communication such as a telephone call. The delegation may be due to the Project Manager being absent for a period due to holidays or illness, or because he wishes to appoint someone to assist him in his duties.

Although the contract is not specific, the notice should identify who the Project Manager is delegating to, what their authority is, and how long the delegation will last, so the Contractor is in no doubt as to who has authority under the contract.

It is important to recognise that in delegating the Project Manager is sharing an authority with the delegate who will then represent him; he is not passing responsibility on to them. The authority is shared, but ultimately responsibility will remain with the Project Manager as the principal. It is also important to note that one can only delegate outwards from the principal, i.e. no one can assume authority possibly because, within an organisation, they are senior to the Project Manager. If the Employer wishes to replace the Project Manager or the Supervisor he must first notify the Contractor of the name of the replacement before doing so.

Question 0.13 What is the meaning of the wording in Clause 10.1 of the NEC3 contracts, the parties named "shall act as stated in the contract and in a spirit of mutual trust and co-operation"?

The first clause of all the NEC3 contracts requires the parties to act in a spirit of mutual trust and co-operation, the Engineering and Construction Contract for example, requiring the Employer, the Contractor, the Project Manager and the Supervisor to do so.

This mirrors Sir Michael Latham in his report "Constructing the Team" when he recommended that the most effective form of contract should include "a specific duty for all parties to deal fairly with each other and in an atmosphere of mutual co-operation".

It is worth considering this obligation in a little detail, as the clause has often been viewed with some confusion, and for those who have spent many years in the construction industry, with a great degree of scepticism. The first part "the Employer, the Contractor, the Project Manager and the Supervisor shall acted as stated in the contract" is straightforward, and many would say is not required to be expressly stated as the named parties have certain obligations within the contract, even though the Project Manager and the Supervisor are not the contracting parties, but the second part "and in a spirit of mutual trust and co-operation" has caused some debate.

Most practitioners state that their understanding of the second part of this clause is that the parties should be non-adversarial towards each other, acting in a collaborative way and working for each other rather than against each other, and in reality that is what the clause requires, and in reality is how all parties to all contracts should behave.

However, the difficulty is that if a party does not act in a spirit of mutual trust and co-operation what can another party who may be affected do? The answer is that the clause is almost unenforceable as it is virtually impossible to define and quantify the breach, or the ensuing damages that flow from the breach. In addition the Contractor is not contractually related to the Project Manager or Supervisor, so either would be unable to take direct action against the other for breach of contract other than through the Employer, who would ultimately be liable for his Project Manager and/or Supervisor complying with the contract.

To that end, Employers have been seen to insert a Z clause to delete Clause 10.1; however, it is strongly recommended that it should remain in the contract, if merely viewed as a statement of good intent. Cynics may say if you delete the clause the parties are not required to act in a spirit of mutual trust and co-operation, but that view can only remain in the domain of cynics!

In effect, a clause requiring parties to act in a certain spirit will probably not, on its own, have any real effect. Within the Engineering and Construction Contract, it is the clauses that follow within the contract which require early warnings, clearly detailed programmes which are submitted for acceptance, and a structured change management process, that actually create and develop that level of mutual trust and co-operation rather than simply inserting a statement within the contract requiring the parties to do so.

Question 0.14 What is the order of precedence of documents in an NEC3 contract?

The NEC3 contracts, unlike most other construction contracts, do not provide a priority or hierarchy of documents within the contract clauses.

However, upon thorough examination of the contract, one can establish that there is a hierarchy between various documents.

For example, if the Contractor has design responsibility and there is an inconsistency between the Works Information provided by the Employer and the Works Information provided by the Contractor, which takes precedence?

The answer can be found in Clauses 11.2(5) and 60.1(1).

Defects are defined in Clause 11.2(5) as:

- a part of the works which is not in accordance with the Works Information, or
- a part of the works designed by the Contractor which is not in accordance with the applicable law or the Contractor's design which the Project Manager has accepted.

Clause 60.1(1): The Project Manager gives an instruction changing the Works Information except:

- a change made in order to accept a Defect, or
- a change to the Works Information provided by the Contractor for his design which is made either at his request or to comply with other Works Information provided by the Employer.

Acceptance of design

The Contractor is required to submit the particulars of his design, as the Works Information requires, to the Project Manager for acceptance. The particulars of the design in terms of drawings, specification and other details must clearly be sufficient for the Project Manager to make the decision as to whether the particulars comply with the Works Information and also, if relevant, the applicable law.

The Works Information may stipulate whether the design may be submitted in parts, and also how long the Project Manager requires to accept the design. Note that in the absence of a stated timescale for acceptance of design, the "period for reply" will apply. Many Project Managers are concerned that they have to "approve" the Contractor's design and therefore they are concerned as to whether they would be qualified, experienced and insured to be able to do so.

Note the use of the word "acceptance" as distinct from "approval". Acceptance denotes compliance with the Works Information or the applicable law. It does not denote that the design will work, that it will be approved by regulating authorities, or that it will fulfil all the obligations which the contract and the law impose; therefore, performance requirements such as structural strength, insulation qualities, etc., do not need to be considered.

A reason for the Project Manager not accepting the Contractor's design is that it does not comply with the Works Information or the applicable law. The Contractor cannot proceed with the relevant work until the Project Manager has accepted the design. Note that under Clause 14.1, the Project Manager's acceptance does not change the Contractor's responsibility to Provide the Works or his liability for his design. This is an important aspect as, if the Project Manager accepts the Contractor's design but following acceptance there are problems with the proposed design – for example, in meeting the appropriate legislation or the requirements of external regulatory bodies – this is the Contractor's liability.

Under Clause 27.1, the Contractor obtains approval of his design from Others where necessary. This will include planning authorities, and other third party regulatory bodies. Note also Clause 60.1(1) second bullet point, which clearly refers to the fact that the Employer's Works Information prevails over the Contractor's Works Information. This clause gives precedence to the Works Information in Part 1 of the Contract Data over the Works Information in Part 2 of the Contract Data. Thus the Contractor should ensure that the Works Information he prepares and submits with his tender as Part 2 of the Contract Data complies with the requirements of the Works Information in Part 1 of the Contract Data.

Question 0.15 What does the term "Working Areas" mean in an NEC3 Engineering and Construction Contract? Presumably this is simply the Site?

The term "Working Areas" is assumed by NEC3 users to mean "the Site", but it is actually a far wider definition than "the Site" and needs a little explanation as confusion often arises.

The contract defines the Working Areas under Clause 11.2(18) as those parts of the working areas which are:

- necessary for Providing the Works, and
- used only for work in this contract.

The Working Areas are initially the Site, the boundaries of which are defined in Contract Data Part 1, but also any additional Working Areas may be identified by the Contractor in Contract Data Part 2 and submitted as part of his tender. An example of an addition to the Working Areas could be where the footprint of the project to be built fills the Site, in which case the Contractor may name an additional area such as an adjacent field which he proposes to use for storage or to locate his Site compound.

In doing so, in order to qualify as an addition to the Working Areas, the Contractor should comply with Clause 11.2(18) in that this additional area is necessary for Providing the Works and is used only for work in this contract. Under Clause 15.1, the Contractor may also submit a proposal to the Project Manager for adding to the Working Areas during the carrying out of the works. The Project Manager may refuse acceptance because the Contractor has not complied with Clause 11.2(18).

The implications of adding to the Working Areas depend on the Main Option chosen.

Options A and B

The Shorter Schedule of Cost Components refers to the cost of resources used within the Working Areas, so these resources if also used within the extended Working Areas would be included as cost rather than within the Fee percentage when assessing compensation events.

Options C, D and E

The Schedule of Cost Components refers to the cost of resources used within the Working Areas, so these resources if also used within the extended Working Areas would be included as cost rather than within the Fee percentage when assessing compensation events.

Question 0.16 How do the NEC3 contracts deal with verbal communications? What do the words "in a form which can be read, copied and recorded" mean?

The NEC contracts have strict rules regarding communications under the contract – for example, Clause 13.1 of the Engineering and Construction Contract states:

> Each instruction, certificate, submission, proposal, record, acceptance, notification, reply and other communication which this contract requires must be communicated in a form which can be read, copied and recorded.

The terms "read, copied and recorded" can include communications sent by electronic means – for example, email or via a project intranet – so the issue and receipt are simultaneous.

Normally the parties agree a protocol for project communications.

This requirement is particularly important in respect of instructions as the contract does not recognise verbal instructions and, whereas under other forms of contract a verbal instruction can be confirmed by the Contractor and if not dissented from, normally within seven or 14 days it becomes an instruction, the NEC3 contracts do not have that provision.

The instruction may be in the form of a letter, a pro forma instruction or an email.

The Project Manager, Supervisor or Contractor replies within the period for reply stated in Contract Data Part 1, unless the contract states otherwise.

This period can be extended by mutual agreement between the communicating parties.

If the Project Manager is required to accept or not accept, he must state his reasons for non-acceptance. Withholding acceptance for a reason stated in the contract is not a compensation event, particularly in respect of:

1 acceptance of Contractor's design
2 acceptance of a proposed Subcontractor
3 acceptance of a Contractor's programme.

The Project Manager issues his certificates (Payment, Completion and Termination Certificates) to the Employer and the Contractor; the Supervisor issues the Defects Certificate to the Project Manager and the Contractor.

Notifications should be communicated separately from other communications, so for example, an early warning notice should not be included as part of a set of meeting minutes, or a letter which includes other subjects.

Question 0.17 We wish to include named specialist Subcontractors in an NEC3 contract. How can we do that?

When Employers are considering including and naming specialist Subcontractors in contracts they can do this in a number of ways:

- nominate them
- novate them to the Contractor once the Contractor is appointed.

Nomination

Although many other contracts provide for the inclusion of Prime Cost Sums within the tender documents in order that Nominated Subcontractors, chosen by the Employer or his representatives, can be brought into the contract later by instructing the Contractor to use them, the NEC contracts have never done so. This is for a number of reasons, as outlined below:

1 The Contractor should be responsible for managing all that he has contracted to do. Nomination will often split responsibilities.
2 The use of Prime Cost Sums in Bills of Quantities means that the Contractor does not have to price that element of the work other than for profit and attendances; the more Prime Cost Sums, the less the tenderers have to price and thus there is less pricing competition between the tenderers.
3 Contracts will often give the Contractor relief in the form of an extension of time in the event of a default by the Nominated

Subcontractor, provided that Contractor has done all that would be reasonable or practicable to manage the default.

4 Contracts will often give the Contractor relief in the form of loss and expense where it cannot be recovered from the Nominated Subcontractor, again provided the Contractor has done all that would be reasonable or practicable to manage the default. This would include the insolvency of a Nominated Subcontractor and a subsequent renomination.

Novation

The principle of novation is probably best known where it applies to design consultants being novated under design and build contracts.

In this case, the Employer preparers his requirements and sends them to the tendering Contractors and they prepare their proposals to match the Employer's requirements; in reality in nearly half of design and build contracts the Employer has already appointed a design team which prepares feasibility proposals and initial design proposals before tenders are invited.

Outline planning permission and sometimes detailed permission may have also been obtained for the scheme before the Contractor is appointed. Each Contractor then tenders on the basis that the Employer's design team will be novated or transferred to the successful tendering Contractor who will then be responsible for appointing the team and completing the design under a new agreement.

This process is often referred to as "novation", which means "replace" or "substitute", and is a mechanism where one party transfers all its obligations and benefits under a contract to a third party. The third party effectively replaces the original party as a party to that contract, so the Contractor is in the same position as if he had been the Employer from the commencement of the original contract. Many prefer to use the term "consultant switch" where the design consultant "switches" to work for the Contractor under different terms as a more accurate definition.

This approach allows the Employer and his advisers time to develop their thoughts and requirements, consider planning consent issues, then, when the design is fairly well advanced, the designers (or, in this case, the Subcontractors) can then be passed to the successful design and build Contractor. The NEC3 contracts do not include pro forma novation agreements, but it is critical that the wording of these agreements be carefully considered, as there are many badly drafted agreements in existence.

Novation can be by a signed agreement or by deed. As with all contracts, there must be consideration, which is usually assumed to be the discharge of the original contract and the original parties' contractual obligations to each other. If the consideration is unclear, or where there is none, the novation agreement should be executed as a deed.

In many cases the agreement states briefly (and very badly) that from the date of the execution of the novation agreement, the Contractor will take the place of the Employer as he had employed the designers or Subcontractors from the beginning, the document stating that in place of the word "Employer" one should read "Contractor"; however, the issues of design liability, inspection, guarantees and warranties may not always apply on a back-to-back basis, so one must take care to draft the agreement in sufficient detail and refer to the correct parties.

In practice, the best way is to have an agreement drafted between the Employer and the Designer or Subcontractor covering the pre-novation period, and a totally separate agreement drafted between the Contractor and the Designer or Subcontractor for the post-novation period.

When a contract is novated, the other (original) contracting party must be left in the same position as he was in prior to the novation being made. Essentially, a novation requires the agreement of all three parties.

As this book is intended as a guide for NEC3 practitioners worldwide, it is not intended to examine specific legal cases within the UK, as they may not apply on an international basis, but there is certainly case law available in terms of design errors in the pre-novation phase being carried through as the liability of the Contractor in the post novation stage, and also the accuracy and validity of site investigations and the Employer's and the Contractor's obligations in terms of checking and taking ownership of such information.

Question 0.18 What happens under an NEC3 Engineering and Construction Contract Option B (priced contract with bill of quantities) if something is clearly indicated on the tender drawings, but is missing from the bill of quantities?

It is a relatively simple question. Let us say you are tendering to build a new sports stadium under the NEC3 Engineering and Construction Contract Option B (priced contract with bill of quantities), and the drawings clearly show a roof. However, the Bill of Quantities make no reference to it. If you notice this at tender stage, would you or should you programme or price the roof? I accept that this is a pretty extreme example, and unlikely to happen so drastically, but smaller examples of this occur on most projects. The initial answer of course is to pick up the telephone during the tender and get clarification – this at the very least will help your credibility and relationship with the potential Employer.

However, what happens for the smaller elements that you genuinely do not notice and hence do not question at tender stage? There are a few contractual facts that we can deal with here, some of which have been subtly changed in NEC3 contracts compared to NEC2.

The Bill of Quantities is not Works Information, it is a contract document.

By the very nature of Option B, the risk in producing and verifying the bill of quantities lies with the Employer (as opposed to Option A, where the risk in missing something off the activity schedule lies with the Contractor).

The Bill of Quantities has not included an element to price specifically for the roof. Clause 20.1 requires the Contractor to provide the works in accordance with the Works Information. Under clause 17.1 this is an ambiguity or inconsistency between contract documents and the Project Manager should give an instruction to resolve the matter:

- Clause 60.6 states that the Project Manager corrects mistakes in the Bill of Quantities which are due to ambiguities or inconsistencies; it also states that each such event is a compensation event.
- Clause 60.7 states that in assessing a compensation event that results from correction of an inconsistency between the Bill of Quantities and another contract document, the Contractor is assumed to have taken the Bill of Quantities as correct.

Taking the above points into account, the Contractor in our example does not need either to price or allow time within the programme to build a roof. The roof itself (if required) will be assessed as a compensation event, and this will assess both the entitlement of time and cost. This leads to the following important conclusions.

As soon as tenderers notice something ambiguous or inconsistent, they should let the Employer know. Tenderers may think they are giving up a potential commercial advantage, but it will give them real credibility during the tender and demonstrate their commitment to working in a "spirit of mutual trust and co-operation", as well as avoiding a contractual or political dispute on the project.

Notwithstanding the above, tenderers should price and programme the Bill of Quantities, not the Works Information.

Any omission from the Bill of Quantities will be assessed as a compensation event and Contractors are entitled to costs of the change as well as the effects of time if the planned completion date is delayed.

Question 0.19 In an NEC3 Engineering and Construction Contract Option E (cost reimbursable contract), how are insurance premiums recovered by the Contractor?

Many Engineering and Construction Contract users believe that the cost of insurance premiums is included as Defined Cost within the Schedules of Cost Components, but this is not true.

The Schedule of Cost Components and the Shorter Schedule of Cost Components both state the following under Item 7.

The following are deducted from cost:

- the cost of events for which this contract requires the Contractor to insure, and
- other costs paid to the Contractor by insurers.

Therefore:

- in the first bullet above, if a cost is incurred for which the Contractor is or was required to insure, then the Contractor cannot recover it as Defined Cost as it is deducted from Defined Cost.
- in the second bullet above, if the Contractor receives payments from the insurers, then these costs are deducted from Defined Cost.

So, how and where are the insurance premiums covered within the contract? Within the Fee percentage.

This will also apply to insurance-based costs such as the cost of performance bonds (Option X13) or parent company guarantees (Option X4).

Question 0.20 We wish to include for price fluctuations within an NEC3 Engineering and Construction Contract. How can we do that?

Option X1: Price adjustment for inflation (used only with Options A, B, C and D). The Employer should make the decision at the time of preparing the tender documents as to whether inflation for the duration of the contract is to be:

- the Contractor's risk – in which case he *should not* select Option X1, or
- the Employer's risk – in which case, he *should* select Option X1.

The default within the Engineering and Construction Contract is that the contract is "fixed price" in terms of inflation, i.e. the Contractor has priced the work to include any inflation he may encounter during the period of carrying out the contract. If Option X1 is chosen, the Prices are adjusted for inflation as the work progresses, by means of a formula.

Note that under Options C and D, whether Option X1 is selected or not, the Price for Work Done to Date is the current cost at the time incurred. Option X1 is then added to the Total of the Prices (the target). Option X1 is not applicable to Options E and F, as the Employer again pays Defined Cost at the time it is incurred. The key components of the formula are:

- the "Base Date Index" (B) is the latest before the Base Date
- the "Latest Index" (L) is the latest available index before the assessment date of an amount due
- the "Price Adjustment Factor" is the total of the products of each of the proportions stated in the Contract data multiplied by (L–B)/B for the index linked to it. Under Options A and B, the amount due includes an amount for price adjustment which is the sum of:
  - the change in the Price for Work Done to Date since the last assessment of the amount due multiplied by the Price Adjustment Factor for the date of the current assessment
  - the amount for price adjustment included in the previous amount due, and
  - correcting amounts, not included elsewhere, which arise from changes to indices used for assessing previous amounts for price adjustment.
  - Example (for simplicity ignoring correcting amounts): The change in the Price for Work Done to Date = £50,000. The Base Date Index (B) = 280.0. The Latest Index (L) = 295.5. The Price Adjustment Factor is therefore (L–B)/B = (295.45–280.0)/280.0 = 0.055. Inflation since the base date is therefore 5.5%. The amount due is therefore £50,000 × 0.055 = £2,750.

- Under Options C and D, the amount due includes an amount for price adjustment which is the sum of:
  - the change in the Price for Work Done to Date since the last assessment of the amount due multiplied by (PAF/(1+PAF)) where PAF is the Price Adjustment Factor for the date of the current assessment, and
  - correcting amounts, not included elsewhere, which arise from changes to indices used for assessing previous amounts for price adjustment.

This amount is then added to the Total of the Prices, i.e. the target. Note that if Option X1 is chosen, then Defined Cost for compensation events is assessed by adjusting current Defined Cost back to the base date.

Option X1 is included within:

- Engineering and Construction Contract
- Professional Services Contract
- Term Service Contract.

Question 0.21 A new legal requirement has just come into force. The Contractor on an NEC3 Engineering and Construction Contract Option C (target contract with activity schedule) has stated that he is entitled to additional payment to recover this. Is he correct?

As with Option X1, the default is that the contract is "fixed price" in terms of changes in the law, i.e. the Contractor has priced the work to include any changes in the law he may encounter during the period of the contract.

If Option X2 is chosen, and a change in the law occurs after the Contract Date, the Project Manager notifies the Contractor of a compensation event. The Prices may be increased or reduced in addition to providing for any delay to Completion.

Note that Option X2 refers to a change in the law of the country in which the Site is located. So, for example, a change in the law in another country where goods are being fabricated for delivery to the Site will not be a compensation event.

Option X2 is included within:

- Engineering and Construction Contract
- Professional Services Contract
- Term Service Contract.

Question 0.22 We are entering into a partnering arrangement with the Contractor within an NEC3 Engineering and Construction Contract. Is there anything we need to do in order to achieve this?

The answer to the question really depends on what you mean by partnering.

The past 20 years, particularly since the publication of the Latham and Egan reports, have seen the growth of partnering and framework agreements.

The US Construction Industry Institute has defined partnering as:

"A long term commitment between two or more organizations for the purpose of achieving specific business objectives by maximizing the effectiveness of each participant's resources ... The relationship is based upon trust, dedication to common goals and an understanding of each other's individual expectations and values."

Partnering is a medium- to long-term relationship between contracting parties, whereby the Contractor, and in turn Consultants and various other parties, are not required to tender competitively on price for each project, but are awarded the work on the basis of their ability to deliver and normally by price negotiation.

The construction industry is known to be a high risk business, and many projects can suffer unexpected cost and time overruns frequently resulting in disputes between the parties. The risks within a project are initially owned by the Employer, who may choose to adopt a "risk transfer" approach where the risks are assigned through the contract to the Contractor who has the opportunity to price and programme for them, or a "risk embrace" approach where the Employer retains the risks. In reality, most contracts are a combination of the two.

The traditional approach to risk management is that of risk transfer, which is fine if the scope of work is clear and well defined; however, in recent years Employers have become more aware that they can achieve their objectives better by adopting a more "old fashioned" risk embrace culture.

The advantages of partnering agreements are as follows:

- The procurement process from conception to commencement on Site, and subsequent completion, can be significantly reduced in terms of time and resources.
- There is no re-tendering time and cost for future projects.
- Relationships can be developed based on trust.

- Contractors are appointed earlier and can contribute to the design and procurement process through ECI (Early Contractor Involvement).
- There tends to be greater cost certainty.
- Continuous improvement can be achieved by transferring learning from one project to the next.
- Better working relationships can be developed, as the parties know each other.
- Prices can be more competitive and resources used more efficiently by continuous flows of work.

The disadvantages of partnering agreements are listed below:

- Obligations and liabilities can become less clear in time as the parties can lapse into informal working patterns.
- Complacency can set in after some time as the Contractor has an assured flow of work.
- Employers are often unsatisfied that they are getting value for money in their projects.
- Continuing to award the work to a small number, or even one Contractor, prevents other equally as good, or better, Contractors having the opportunity to carry out work.

All of these disadvantages can be overcome by maintaining a disciplined approach to communications between the parties and their rights and obligations, and also by continuous measurement of performance and deliverables.

Secondary Option X12 enables a multi-party partnering agreement to be implemented. In this case Option X12 is used as a Secondary Option common to the contract which each party has with the body which is paying for the work. The parties together make up the partnering team.

The content of Option X12 is derived from the "Guide to Project Team Partnering" published by the Construction Industry Council (CIC).

It is estimated that Option X12 is used on 9 per cent of NEC3 contracts (RICS Contract in Use Survey). It must be stressed that no legal entity is created between the Partners, so it is not a partnership as such. Some definitions need to be explained:

1 "The Partners" are those named in the Schedule of Partners.
2 An "Own Contract" is a contract between two Partners which includes Option X12.

3 The "Core Group" comprises the Partners listed in the Schedule of Core Group Members.
4 "Partnering Information" is information which specifies how the Partners work together.
5 A "Key Performance Indicator" is an aspect of performance for which a target is attached in the Schedule of Partners. Each Partner, represented by a single individual, is required to work with the other Partners in accordance with the Partnering Information to achieve the Employer's objective stated in the Contract Data and the objectives of every other Partner.

The Core Group, led by the Employer's Representative, its members selected by the Partners, acts and takes decisions on behalf of the

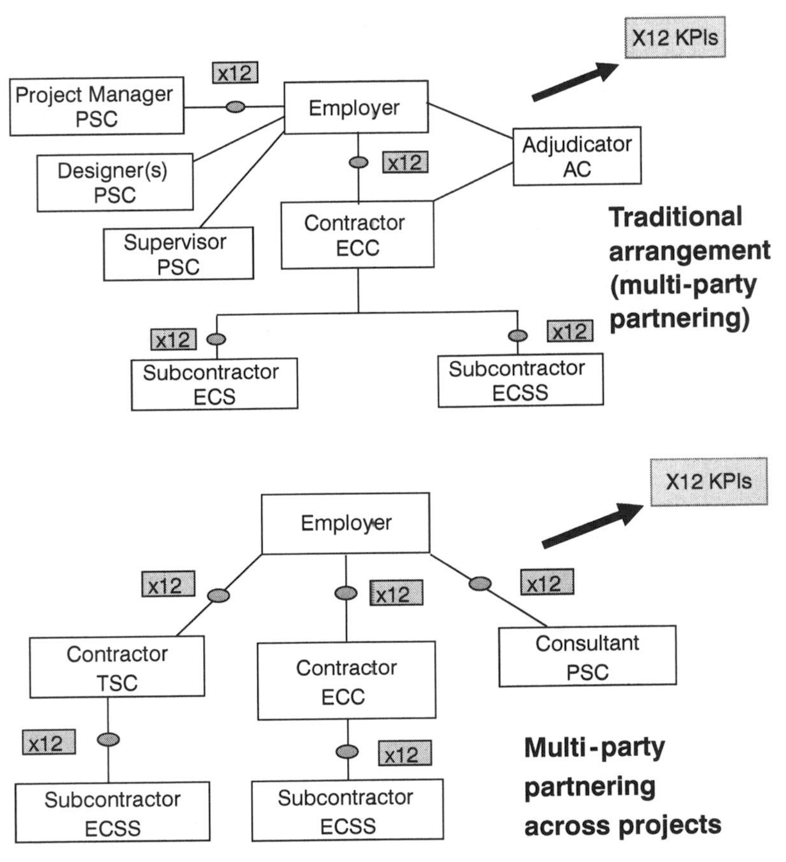

Figure 0.4 Use of Option X12

Partners. The Core Group also keeps up to date a Schedule of Core Group Members and a Schedule of Partners.

The Partners are required to work together, using common information systems, and a Partner may ask another Partner to provide information which he needs to carry out the work in his Own Contract. Each Partner gives an early warning to the other Partners when he becomes aware of anything that could affect another Partner's objectives.

The Core Group may give an instruction to the Partners to change the Partnering Information, which is a compensation event. The Core Group also maintains a timetable showing the Partners' contributions. If the Contractor needs to change his programme to comply with the timetable, then it is a compensation event. Each Partner also gives advice, information and opinion to the Core Group where required. Each Partner must also notify the Core Group before subcontracting any work, though it does not say that the Core Group is required to respond to the notification.

Finally, Option X12 provides for Key Performance Indicators (KPIs) with amounts paid as stated in the Schedule of Partners. The Employer may add a KPI to the Schedule of Partners but cannot delete or reduce a payment.

Option X12 is included within:

- Engineering and Construction Contract
- Professional Services Contract
- Term Service Contract.

Question 0.23 Within the NEC3 Engineering and Construction Contract, when would we use Option X15 (limitation of the Contractor's liability for his design to reasonable skill and care)?

Whilst the Works Information defines what, if any, design is to be carried out by the Contractor, the contract is silent on the standard of care to be exercised by the Contractor when carrying out any design. Two terms that relate to design liability are "fitness for purpose" and "reasonable skill and care".

Whilst this book is intended for international use, in defining the term "fitness for purpose" one must look in English law to the Sale of Goods Act 1979, which refers to the quality of goods supplied including their state and condition complying in terms of "fitness for all the purposes for which goods of the kind in question are commonly supplied".

In construction, fitness for purpose means producing a finished project fit in all respects for its intended purpose. This is an absolute duty independent of negligence, and in the absence of any express terms within the contract to the contrary, a Contractor who has a design responsibility will be required to design and build the project "fit for purpose".

Some contracts will limit the Contractor's liability to that of a consultant, i.e. reasonable skill and care. The Engineering and Construction Contract does this through Option X15. If this Secondary

Example

The Works Information requires the Contractor to design and install an HVAC system to a major retail development. There are specific and measurable performance criteria for the system including temperature variations and energy efficiency. The system is required to have a design life of 30 years (360 months).

The Works Information states that the system will be tested at Completion and includes a table showing how the performance of the system will be measured and acceptable levels of achievement.

The system is expected to perform to 98–100% of performance criteria.

If it falls within 90–98% the system will be accepted, but low performance damages will be payable by the Contractor to the Employer.

If it falls below 90% it will not be accepted.

Contract Data Part 1 contains the following entries:

Amount (per month)	*Performance per month level*
£150.00	94% – 96%
£200.00	92% – 94%
£250.00	90% – 92%

When tested, the system achieves 95% of the performance criteria.

Therefore,

360 months × £150.00 = £54,000

£54,000 is payable by the Contractor at the Defects Date. Contrary to popular belief, the Employer does not collect the damages from the Contractor every month!

Option is not chosen, the Contractor's liability for design is "fitness for purpose".

Contrast this with the level of liability of a consultant (whether acting for an Employer or a Contractor) in providing a design service, which is defined, again in English law by the Supply of Goods and Services Act 1982, where there is an implied term that the consultants will carry out the service – in this case design – with reasonable care and skill, which means designing to the level of an ordinary, but competent person exercising a particular skill.

So, in the absence of any express terms to the contrary, a designer will normally be required to design using "reasonable skill and care". This is normally achieved by the designer following accepted practice and complying with national standards, codes of practice, etc.

Clearly, despite the Contractor believing that he has offset his design obligations and liability to his designing consultant, he has to be aware that he and his consultant have differing levels of care and liability.

Question 0.24 We note that the NEC3 Engineering and Construction Contract includes Option X17 (low performance damages). What is this, and how is it calculated and paid to the Employer?

In the event that the Contractor produces defective work, the Employer has three options:

- The Contractor corrects the Defect (Clause 43.1).
- If the Contractor does not correct the Defect, the Project Manager assesses the cost to the Employer of having the Defect corrected by other people and the Contractor pays this amount (Clause 45.1).
- The Employer can accept the Defect and a quotation from the Contractor for reduced Prices and/or an earlier Completion Date (Clause 44).

Where the performance of the works or an element of the works fails to reach the specified level within the contract, the Employer can take action against the Contractor to recover any damages suffered as a result of the breach, but as an alternative can recover low performance damages under Option X17 if it has been selected.

Question 0.25 We do not understand NEC3 Engineering and Construction Option X18 (limitation of liability). When would this be applied?

If the Contractor causes any loss of, or damage to, the Employer's property, he would normally be liable for the full cost of remedial works; however, Clause X18 provides for this liability to be limited to amounts stated in the Contract Data. In addition, the Contractor's liability to the Employer for latent defects due to his design may again be limited to amounts stated in the Contract Data. Clause X18.4 can be used to place limits on the total liability the Contractor has to the Employer for all matters under the contract other than excluded matters in contract, tort or delict. Excluded matters are:

- loss or damage to Employer's property
- delay damages if Option X7 applies
- low performance damages if Option X17 applies
- Contractor's share if Option C or D applies.

The Contractor is not liable for any matter unless it has been notified to the Contractor before the end of liability date which is stated in the Contract Data in terms of years after the Completion of the whole of the works. In the UK, this may be 6 or 12 years dependent on the type of contract; other legislations set this at 10 years.

Question 0.26 How can we include for Key Performance Requirements (KPIs) with an NEC3 Engineering and Construction Contract?

Performance of the Contractor can be monitored and measured against Key Performance Indicators (KPIs) using Option X20. Targets may be stated for Key Performance Indicators in the Incentive Schedule. The Contractor is required to report his performance against KPIs to the Project Manager at intervals stated in the Contract Data including the forecast final measurement. If the forecast final measurement will not achieve the target stated in the Incentive Schedule, the Contractor is required to submit his proposals to the Project Manager for improving performance.

The Contractor is paid the amount stated in the Incentive Schedule if the target for a KPI is improved upon or achieved. Note that there is no payment due from the Contractor if he fails to achieve a stated target. The Employer may add a new KPI and associated payment to the Incentive Schedule but may not delete or reduce a payment.

Option X20 is included within:

- Engineering and Construction Contract
- Professional Services Contract
- Term Service Contract.

Question 0.27 How should Z clauses be incorporated into an NEC3 contract? Are there any recommended Z Clauses?

Probably the best answer to this question is that Z clauses should be used "as sparingly as possible"!

Option Z allows conditions to be added to, or omitted from, the core clauses.

All changes to the core clauses should be included as Z clauses rather than amending the core clauses themselves, so in effect the clause remains in the contract but is amended within the Z clause. It is also critical that when drafting a Z clause, it must be clearly stated what happens to the original core clause – for example, it is deleted. If the original core clause is not deleted, then it is likely than an inconsistency will arise which will be interpreted against the party who wrote or amended the clause, i.e. the Employer.

These conditions may modify or add to the core clause to suit any risk allocation or other special requirements of the particular contract. However, changes should be kept to a minimum, consistent with the objective of using industry standard, impartially written contracts.

It must be remembered that if an Employer amends a contract to allocate a risk to the Contractor which may have been intended to be held by the Employer, the Contractor has a right to price it in terms of time and money; therefore, the practice of Employers amending contracts to pass risk to Contractors without considering who is best able to price, control and manage those risks can, in many cases, prove to be unwise and uneconomical.

It is important to spend time considering whether a Z clause is appropriate in each case, then when that decision has been made, that the clause is drafted correctly and aligned to the drafting principles of the original contract, in the case of the Engineering and Construction Contract using ordinary language, present tense, short sentences, bullet pointing and italicizing of terms identified in the Contract Data. It is not unusual to see Z clauses in an Engineering and Construction Contract written in a legalistic language, in the future tense, with no italics, and without punctuation apart from full stops!

There are no recommended Z clauses, but some which tend to be used fairly commonly are as follows.

Payment

Assessing the amount due 50

Delete: Clause 50.4 – In assessing the amount due, the *Project Manager* considers the application for payment the *Contractor* has submitted on or before the assessment date. The *Project Manager* gives the *Contractor* details of how the amount due has been assessed.

Add: The *Contractor* submits an application for payment one week before each assessment date. In assessing the amount due, the *Project Manager* considers the application for payment the *Contractor* submits. The *Project Manager* gives the *Contractor* details of how the amount due has been assessed.

Reason for Z clause: There is no obligation for the Contractor to submit an application for payment; the Project Manager has to assess the payment due to the Contractor whether or not an application has been submitted. For that reason, users tend to make the Contractor submit an application as follows:

The Contractor's share

Delete: Clause 53.3 – The *Project Manager* makes a preliminary assessment of the *Contractor's* share at Completion of the whole of the *works* using his forecasts of the final Price for Work Done to Date and the final total of the Prices. This share is included in the amount due following Completion of the whole of the *works*.

Add: The *Project Manager* makes a preliminary assessment of the *Contractor's* share at any assessment date. If the forecast Price for Work Done to Date is less than the final total of the Prices, this share is included in the amount due following Completion of the whole of the *works*.

If the forecast Price for Work Done to Date is more than the forecast total of the Prices, this share is retained from the amount due.

Reason for Z clause: Many Employers believe that the Contractor's share should be calculated when the Price for Work Done to Date reaches the current Total of the Prices (the target) rather than on

Completion, so that the Contractor is not paid monies that he will have to pay back later.

For Options C, D, E and F

Delete: Clause 11.2(29) – The Price for Work Done to Date is the total Defined Cost which the *Project Manager* forecasts will have been paid by the *Contractor* before the next assessment date plus the Fee.

Add: The Price for Work Done to Date is the total Defined Cost which the Contractor has paid plus the Fee.

Reason for Z clause: Many Employers do not accept Clause 11.2(29) in that a forecast has to be made of the Defined Cost up to the *next* assessment date.

Testing and Defects

Searching for and (42) notifying Defects

Delete: Clause 42.2 – Until the *defects date*, the *Supervisor* notifies the *Contractor* of each Defect as soon as he finds it and the *Contractor* notifies the *Supervisor* of each Defect as soon as he finds it.

Add: Until the *defects date*, the *Supervisor* notifies the *Contractor* of each Defect as soon as he finds it.

Reason for Z clause: Why does the Supervisor need to know the Contractor has found a Defect?

Compensation Events

Assessing compensation events 63

Reason for Z clause: The default position with pricing compensation events is that they are assessed as the effect of the compensation event upon actual/forecast Defined Cost plus the resulting Fee (Clause 63.1), *but* under Option A, C, E and F (Clause 63.14) and Option B and D (Clause 63.13) if the Project Manager and Contractor agree rates and lump sum prices may be used instead.

These clauses are sometimes *reversed* so that compensation events are based on agreed rates and prices – e.g. bill rates, etc. – *unless* the Project Manager and Contractor agree to use Defined Cost plus the resulting Fee.

Question 0.28 We are confused by the fact that within the Engineering and Construction Contract are the Schedule of Cost Components and the Shorter Schedule of Cost Components. What are these for and what is the difference between the two?

Later in this answer we will examine in detail the Schedule of Cost Components and the Shorter Schedule of Cost Components. First, it is important to say that the two Schedules of Cost Components have two uses:

- Under Options A, B, C, D and E, they define the cost components, which are included in assessment of compensation events.
- Under Options C, D and E, they also define the cost components for which the Contractor will be directly reimbursed.

Note: The Schedules of Cost Components do not apply to Option F (Management Contract) where Defined Cost is the amount of payments to Subcontractors plus the cost of any work carried out by the Contractor himself:

- For Options A and B, the Shorter Schedule of Cost Components is a complete statement of the cost components under the definition of Defined Cost, including Subcontractors.
- For Options C, D and E the Schedule of Cost Components excludes payments due to Subcontractors who have to be added to calculate the Defined Cost.

The "Schedule of Cost Components" or the "Shorter Schedule of Cost Components"?

There are two Schedules of Cost Components in the contract because there are differing uses for Defined Cost, dependent on which Main Option is chosen (see below and Table 0.1):

Option A) Option B)	Defined Cost is only used for assessing compensation events, therefore the Shorter Schedule of Cost Components is used.
Option C) Option D)	Defined Cost is used for assessing compensation, events, and for assessing Price for Work Done to Date.

Option E) Therefore, the of Cost Components is used, though the Shorter Schedule of Cost Components can be used by agreement between the Project Manager and the Contractor for assessing compensation events.

Option F) Defined Cost is the amount of payment due to Subcontractors and the prices for work done by the Contractor himself. Therefore, neither the Schedule of Cost Components nor Shorter Schedule of Cost Components is used.

Table 0.1 Differences between the Schedule of Cost Components and the Shorter Schedule of Cost Components

Schedule of Cost Components	*Short Schedule of Cost Components*
People	
Reference to the Contractor means the Contractor and not his Subcontractors. People costs classified as Defined Cost.	Reference to the Contractor means the Contractor and his Subcontractors. No detailed list, but specifically includes amounts paid by the Contractor for meeting the requirements of the law and for pension provision.
Equipment	
Detailed list of Equipment costs classified as Defined Cost.	Reference to the published list identified within Contract Data Part 2.
Plant and Materials	
No difference	
Charges	
Detailed list of Charges classified as Defined Cost. Overhead costs calculated by reference percentage for Working Areas Overheads.	No detailed list as Defined Cost. Overhead costs calculated by reference to the percentage for People Overheads.
Manufacture and fabrication	
Reference to hours worked multiplied by rates in Contract Data Part 2.	Reference to amounts paid by the Contractor.
Design	
No difference.	
Insurance	
No difference.	

The Schedule of Cost Components in detail

The Schedule of Cost Components only applies when Options C, D or E are used.

References to the Contractor do not include his Subcontractors.

Clause 1: People

This relates to the cost of people who are directly employed by the Contractor and whose normal place of working is within the Working Areas, i.e. the site and any another area named in Contract Data Part 2, and also people whose normal place of working is not within the Working Areas but who are working in the Working Areas, i.e. people who are based off site, but who are on site temporarily.

The cost component covers the full cost of employing the people including wages and salaries paid whilst they are in the Working Areas, payments made to the people for bonuses, overtime, sickness and holiday pay, special allowances, and also payments made in relation to people for travel, subsistence, relocation, medical costs, protective clothing, meeting the requirements of the law, a vehicle and safety training.

NEC3 contracts do not provide for people rates to be priced as part of the Contractor's tender, the principle being that the People component is real cost rather than a rate that has been forecast at tender stage months or even years before it is required to be used.

This is a fairer method of establishing cost in that the risk is not the Contractor's, but in practice requires the Contractor if necessary to prove the cost of each operative, tradesman and member of staff, which can at times be laborious. For this reason, many employers include a schedule of rates to be priced by the Contractor at tender stage, these rates then being used for payments and compensation events where appropriate. These rates are also referred to during the tender assessment process.

Clause 2: Equipment

This relates to the cost of Equipment (referred to as "plant" in other contracts) used within the Working Areas. If the Equipment is used outside the Working Areas, then it is deemed to be included in the fee percentage. If the equipment is hired externally by the Contractor, the hire rate is multiplied by the time the equipment is required.

The cost of transport of Equipment to and from the Working Areas and any erection and dismantling costs are also costed separately. If the Equipment is owned by the Contractor, or hired by the Contractor from a company within the parent company such as an internal "plant hire" company, the cost is assessed at open market rates (not at the rate charged by the hirer) multiplied by the time for which the Equipment is required.

Difficulty has often arisen with past editions of the contract where a piece of equipment is owned by the Contractor, or a company within the group, in which case there would be no invoice as such to prove the cost – in many cases it was just an internal charge or cost transfer.

In those previous editions, the drafters sought to deal with the problem by establishing the weekly cost by the use of a formula:

> (Purchase price of the Equipment/Working life remaining at purchase) × Depreciation and maintenance %

In theory, this should provide an equitable solution, but in practice it has provided some inequitable answers and has to be applied to every piece of equipment owned by the Contractor! One hopes he did not have too many of his own wheelbarrows on the site!

In NEC3 the drafters then changed to "open market rates multiplied by the time for which the equipment is required".

In truth major contractors who have a large "plant" company will benefit in this respect from advantageous discount agreements with their suppliers which are well below "open market rates".

Equipment purchased for use in the contract is paid on the basis of its change in value (the difference between its purchase price and its sale price at the end of the period for which it is used) and the time related on cost charge stated in Contract Data Part 2 for the period the Equipment is required.

During the course of the contract, the Contractor is paid the time related charge per time period (per week/per month) and when the change in value is determined, a final payment is made in the next assessment.

- Example: The Contractor has purchased six Portakabins at £6,250 each for use on a project of four years' duration. This cost includes supply and delivery of the Portakabins:
  - Total purchase price = 6 × £6,250 = £37,500
  - On completion, the Portakabins are sold for £2,750 each; therefore the total sale price is £16,500
  - The change in value is therefore £37,500–£16,500 = £21,000

- Example:
  - Purchase Price = £30,000
  - Working life remaining = 250 weeks
  - Depreciation and maintenance percentage = 20%
  - The weekly cost is: £30,000 × 1.20 = £144.00 per week x 250

Any special Equipment is paid for on the basis of its entry in Contract Data Part 2. Consumables such as fuel are also separately costed, including any materials used to construct or fabricate equipment.

The cost of transporting Equipment to and from the Working Areas, and the erection, dismantling and any modifying of the Equipment is costed separately.

Any People cost such as equipment drivers and operatives involved with erection and dismantling of plant should be included in the cost of People, not the Equipment they work on.

Clause 3: Plant and Materials

This deals with purchasing Plant and Materials, including delivery, providing and removing packaging, and any necessary samples and tests. The cost of disposal of Plant and Materials is credited.

Clause 4: Charges

This covers various miscellaneous costs incurred by the Contractor such as temporary water, gas and electricity, payments to public authorities, and also payments for various other charges such as cancellation charges, buying or leasing of land, inspection certificates and facilities for visits to the Working Areas.

The cost of any consumables and equipment provided by the Contractor for the Project Manager's or Supervisor's offices is also included as direct cost. (Note: The cost of the Contractor's own equipment is covered in the Equipment Section, and the Contractor's consumables are included within the Working Areas overheads percentage.)

Item 44 covers various site consumables as listed from (a) to (j) (see Table 0.2).

It is important to recognise that Item 44 only includes for provision and use of equipment, supplies and services, but excludes accommodation. So if we consider the list, it is important to recognise what *is* and what *is not* included.

Table 0.2 The Schedule of Cost Components

Item 44	*Included*	*Not Included*	*Where is it included*
(a) Catering	Catering equipment Food Electricity Bills	Canteen accommodation Catering staff	Equipment People
(b) Medical facilities	First aid box Stretchers	First aid personnel First aid accommodation	People Accommodation
(c) Recreation	Television Radio Gym Facilities		
(d) Sanitation	Toilet rolls Hand towels Soap Barrier creams	Toilet Accommodation	Equipment
(e) Security	Security systems Alarms	Security personnel Guard dogs	People Equipment
(f) Copying	Photocopier Photocopier toner and paper	Other stationery	Equipment
(g) Telephone, telex, fax, radio and CCTV	Telephones Fax Machine		
(h) Surveying and setting out	Surveying equipment	Site Engineer Surveyors	People
(i) Computing	Hardware Software Printers Print paper	IT Personnel	People
(j) Hand tools not powered by compressed air	Electric tools		

The cost is calculated by multiplying the Working Areas overheads percentage inserted by the Contractor in Contract Data Part 2 by the People cost items 11, 12, 13 and 14.

Clause 5: Manufacture and fabrication

This relates to the components of cost of manufacture or fabrication of Plant and Materials outside the Working Areas. Hourly rates are stated in Contract Data Part 2 for the categories of employees listed.

Clause 6: Design

This deals with the cost of design outside the Working Areas. Again, hourly rates are stated in Contract Data Part 2 for the categories of employees listed.

Clause 7: Insurance

The cost of events for which the Contractor is required to insure and other costs to be paid to the Contractor by insurers are deducted from cost.

The Fee

Whilst many practitioners believe that the Fee simply covers "overheads and profit", the definition is a little wider. All components of cost not listed in the Schedules of Cost Components are covered by the fee percentage.

There is no schedule of items covered by the fee percentage, but the following list, whilst not exhaustive, gives some examples of cost components not included in the Schedules of Cost Components:

1 Profit.
2 The cost of offices outside the Working Areas, e.g. the Contractor's head office.
3 Insurance premiums.
4 Performance bond costs.
5 Corporation tax.
6 Advertising and recruitment costs.
7 Sureties and guarantees required for the contract.
8 Some indirect payments to staff.

From this one can see that there are some elements covered by the Fee percentage which could be mistakenly assumed as being covered by the Schedules of Cost Components.

Previous editions of the Engineering and Construction Contract had only one fee percentage applicable to total cost, but the NEC3 Engineering and Construction Contract allows the Contractor to tender two fee percentages:

- the subcontracted fee percentage applied to the Defined Cost of subcontracted work.
- the direct fee percentage applied to the Defined Cost of other work. It is essential that these two fee percentages are correctly allocated to the appropriate costs when assessing payments and compensation events. In reality, many Contractors tend to bracket the two fee percentages together as a single fee percentage, which they are entitled to do, and this in effect makes life easier for the Contractor and the Project Manager when assessing payments and compensation events.

Note: Under Options A and B, Defined Cost is the cost of components in the Shorter Schedule of Cost Components whether work is subcontracted or not, whereas under Options C, D and E, where the Schedule of Cost Components is normally used, reference to the Contractor means the Contractor and not his Subcontractors. Subcontractors' costs are dealt with separately.

The Shorter Schedule of Cost Components in detail

The Shorter Schedule of Cost Components is restricted to the assessment of compensation events under Option A and B, but if the Project Manager and Contractor agree it may be used for assessing compensation events under Options C, D and E.

Clause 1: People

This relates to the cost of people who are directly employed by the Contractor and whose normal place of working is within the Working Areas, i.e. the site and any another area named in Contract Data Part 2, and also people whose normal place of working is not within the

Working Areas but who are working in the Working Areas, i.e. people who are based off site, but who are on site temporarily.

It also includes people who are not directly employed by the Contractor but are paid according to the time worked whilst they are in the Working Areas, e.g. security guards, cleaners, etc. Whilst there is not a definitive list of costs included as with Items 11, 12 and 13 of the Schedule of Cost Components, it includes amounts paid by the Contractor including those for meeting the requirements of the law and pension provision.

It also includes for the cost of people who are not directly employed by the Contractor but are paid according to the time worked whilst they are in the Working Areas. This may include for example security guards or cleaners who are paid by the hour.

Again, although NEC3 contracts do not provide for people rates to be priced as part of the Contractor's tender, many clients include a schedule of rates to be priced by the Contractor at tender stage, these rates then being used for payments and compensation events where appropriate. These rates are also referred to during the tender assessment process.

Clause 2: Equipment

This relates to the cost of equipment used within the Working Areas. The cost of equipment is calculated by reference to a published list, for example BCIS (Building Cost Information Service) Schedule of Basic Plant Charges or the CECA (Civil Engineering Contractors Association) Dayworks Schedule.

In Contract Data Part 2, the Contractor names the published list to be used and also inserts a percentage for adjustment against items of equipment in the published list. The Contractor additionally inserts rates into Contract Data Part 2 for Equipment not included within the published list.

Any Equipment required which is not in the published list or priced within Contract Data Part 2 is then priced at competitively tendered market rates. The time the Equipment is used is as referred to in the published list, which may have hourly, weekly or monthly rates.

By referring to the published list, whether the Equipment is then owned or hired by the Contractor, is irrelevant. The cost of transporting Equipment to and from the Working Areas and the erection and dismantling of the Equipment is costed separately.

The cost of Equipment operators is included within the People costs. Any Equipment not included within the published lists is priced at competitively tendered or open market rates.

Clause 3: Plant and Materials

This is exactly the same as for the Schedule of Cost Components and also deals with purchasing, delivery, providing and removing packaging, and any necessary samples and tests. The cost of disposal of Plant and Materials is credited.

Clause 4: Charges

This covers various miscellaneous costs incurred by the Contractor such as temporary water, gas and electricity payments to public authorities, and also payments for various other charges, which may or may not be relevant dependent on the project. These costs are not calculated on a direct cost basis but by reference to the percentage for People overheads which is applied to People costs.

Clause 5: Manufacture and fabrication

This relates to the components of cost of manufacture or fabrication of Plant and Materials outside the Working Areas. The calculation is based on amounts paid by the Contractor.

Clause 6: Design

This is exactly the same as for the Schedule of Cost Components and deals with the cost of design outside the Working Areas. Again, hourly rates are stated in Contract Data Part 2 for the categories of employees listed.

Clause 7: Insurance

The cost of events for which the Contractor is required to insure and other costs to be paid to the Contractor by insurers are deducted from cost.

Question 0.29 How are Subcontractors dealt with within the NEC3 Engineering and Construction Contract Schedules of Cost Components?

The NEC3 Engineering and Construction Contract has two Schedules of Cost Components: the Schedule of Cost Components and the Shorter Schedule of Cost Components.

- For Options A and B, the Shorter Schedule of Cost Components is a complete statement of the cost components under the definition of Defined Cost, including Subcontractors.
- For Options C, D and E the Schedule of Cost Components excludes payments due to Subcontractors, which have to be added to calculate the Defined Cost.

See also the previous answer.

Question 0.30 The NEC3 Engineering and Construction Short Contract includes within the Contract Data a Price List. What is this and how do we use it?

Most of the NEC3 contracts include Main and Secondary Options allowing the user to select a procurement strategy and payment mechanism most appropriate to the project and the various risks involved. Essentially the main options differ in the way the Contractor is paid. For example, the Engineering and Construction Contract provides the following Main Options:

Option A	Priced contract with activity schedule
Option B	Priced contract with bill of quantities
Option C	Target contract with activity schedule
Option D	Target contract with bill of quantities
Option E	Cost reimbursable contract
Option F	Management contract.

- Options A and B are priced contracts in which the risks of being able to carry out the work at the agreed prices are largely borne by the Contractor.
- Options C and D are target contracts in which the Employer and Contractor share the financial risks in an agreed proportion.

- Options E and F are two types of cost reimbursable contract in which the financial risks of being able to carry out the work are largely borne by the Employer.

Note: The lettering of the main options is common for all the NEC3 contracts, so Option A is always a Priced Contract, Option C is always a Target Contract, Option E is always cost reimbursable, etc.

For example, the NEC3 Term Service Contract has:

Option A	Priced contract with price list
Option C	Target contract with price list
Option E	Cost reimbursable contract.

As the Engineering and Construction Short Contract is used for "contracts which do not require sophisticated management techniques, comprise straightforward work and impose only low risks on both the Employer and the Contractor", there is no need for a wide range of procurement strategies as provided in the NEC3 Engineering and Construction Contract and other NEC3 contracts, so the Engineering and Construction Short Contract includes within the Contract Data a single page Price List.

The columns are headed Item Number, Description, Unit, Quantity, Rate and Price.

As the guidance note above the Price List states:

- entries in the first four columns are made either by the Employer or the tenderer (Contractor)
- if the Contractor is to be paid an amount for the item which is not adjusted if the quantity of work changes, the tenderer enters the amount in the Price column only, the Unit, Quantity and Rate columns being left blank.

This is then a lump sum contract.

- If the Contractor is to be paid an amount for the item of work which is the rate for the work multiplied by the quantity completed, the tenderer enters the rate which is then multiplied by the expected quantity to produce the Price, which is also entered.

This is then a remeasurement contract.

Target contracts and cost reimbursable contracts are not provided for as they are regarded as unsuitable and too complex for this type of work. Similarly Management Contracts will not be used on the type of work which the Engineering and Construction Short Contract is designed for.

When considering payment to the Contractor using the Price List, one must consider two terms, as follows.

1 **The Prices:** These are the various elements that make up the total Price and are the amounts stated in the Price column of the Price List.
2 **The Price for Work Done to Date:** This term is used in making the assessment of amounts due to the Contractor, and is the total of:
   - the Price for each lump sum item in the Price List which the Contractor has completed, and
   - where a quantity is stated for an item in the Price List, an amount calculated by multiplying the quantity which the Contractor has completed by the rate

   plus
   - other amounts to be paid to the Contractor (including any tax which the law requires the Employer to pay to the Contractor) – this will include VAT

   less
   - amounts to be paid by or retained from the Contractor.

Question 0.31 We note that unlike other contracts we have been using, there does not appear to be within the NEC3 contracts a Form of Agreement for the parties to sign. Why is this?

There is no Standard Form of Agreement for completion and signature within the NEC3 contracts, as the drafters of the contracts have stated that parties enter into contracts in many different ways, by executing Forms of Agreement, by exchange of letters, etc.

Whilst this is probably true, it would have assisted the parties if there had been a standard Form of Agreement that they could use if they so wish, though within the Engineering and Construction Contract Guidance Notes there is a sample form.

The difficulty with not having a Form of Agreement within the Contract is that the Employer has to write his own, which requires them to summarise the documents which comprise the contract and also to consider whether the contract is to be executed as a deed. In that respect they must be very cautious that all the documents intended to form the contract are included in the Form of Agreement.

Having said this, there is however a signature page in the Contract Data of the various Short Contracts within the NEC3 family, for example the Engineering and Construction Short Contract has a page which consists of the Contractor's Offer and the Employer's Acceptance.

The Contractor's Offer

The information the Contractor is required to complete is:

- the contact details of the Contractor
- the percentage for overheads and profit added to the Defined Cost for people
- the percentage for overheads and profit added to other Defined Cost
- offered total of the Prices
- a signature on behalf of the Contractor.

The Employer's Acceptance

The Employer's Acceptance requires a signature by a person authorised to sign on behalf of the Employer.

Question 0.32 Can you explain what "Option X19: Providing the services by Task Order" under the NEC3 Term Service Contract means?

This Option can be used with any of the three main Options within the NEC3 Term Service Contract to enable the Employer to "call off" items of work from the Price List and instruct the Contractor to carry them out.

The instruction takes the form of a Task Order, which describes the work to be done, and the dates within which it is to be done.

It is very useful where the precise quantity and location of work required is not known at the Contract Date.

Items of work expected to be instructed should be included in the Price List as far as possible. No work is done under X19 until a Task Order is issued to the Contractor.

Option X19 defines a Task as "work which the Service Manager may instruct the Contractor to carry out within a stated period of time".

A Task Order is "the Service Manager's instruction to carry out a task".

The Task Order will include:

- a detailed description of the work within the Task
- a priced list of items of work in the Task in which items taken from the Price List are identified
- the starting and completion dates for the Task
- the amount of delay damages for the late completion of the Task, and the total of the Prices for the Task when Option A or C is used, or the forecast total of the Prices for the Task if Option E is used.

Option X19 also requires the Contractor to submit a programme to the Services Manager within the period from recovering the Task Order stated in the Contract Data.

No Task Order is issued after the end of the service period.

Chapter 1

Early warnings and Risk Registers

Question 1.1 Is there a standard format within the NEC3 contracts for an early warning notice? Is there any remedy if the Project Manager or the Contractor fails to give an early warning?

There is no standard format for an early warning notice, but it must contain certain essential ingredients. In order to consider these essentials, let us first consider what an early warning is.

It is important to state that early warnings are an essential and valuable risk management tool within the NEC3 contracts and are included within all the contracts except the Adjudicator's Contract.

If we take the Engineering and Construction Contract as an example, the contract obliges the Project Manager and the Contractor to notify each other as soon as either becomes aware of any matter which could affect the project (the "works") in terms of time, cost or quality.

Under Clause 16.1 the Contractor and the Project Manager give an early warning by notifying the other as soon as either becomes aware of any matter which could:

- increase the total of the Prices (the price of the works, in the form of the activity schedule, the Bill of Quantities or the target)
- delay Completion (the Completion of the whole of the works)
- delay meeting a Key Date (the Completion of an intermediate "milestone date" in accordance with Clause 11.2(9)
- impair the performance of the works in use.

This last point sometimes causes confusion, but if we take as an example the Contractor being instructed by the Project Manager to use a particular type of water pump and the Contractor knows from experience that particular pump would probably not be sufficient to meet the Employer's requirements once the works are taken over, then the Contractor should give early warning (see Figure 1.1).

The Contractor may also give an early warning by notifying the Project Manager of any other matter which could increase his total cost. One could query whether a matter which could increase the Contractor's cost, but not affect the Price, should be an early warning matter, or for that matter, whether it should be anything to do with the Project Manager, particularly if Options A or B have been selected so the Contractor is not looking to recover this additional cost through the contract, but the words are "the Contractor may give an early warning" so he is not obliged to do so.

Note, also within Clause 16.1, the Contractor is not required to give an early warning for which a compensation event has previously been notified. So, as an example, if the Project Manager gives an instruction which changes the Works Information, it is a compensation event, for which neither the Project Manager nor the Contractor are required to give early warning.

The early warning procedure obliges people to be "proactive", dealing with risks as soon as the parties become aware of them, rather than "reactive", waiting to see what effect they have then trying to deal with them when it is often too late.

Encouraging the early identification of problems by both parties puts the emphasis on joint solution finding rather than blame assignment and contractual entitlement.

It is one of the most important and valuable aspects of the contract and it is perhaps surprising that, whilst few other contracts refer to an early warning process, only the NEC contracts set out in clear detail what the parties are obliged to do, with appropriate sanctions should the parties not comply (see Clauses 11.2(25), 61.5 and 63.5).

Notifying early warnings

Again, if we refer back to the question, "Is there a standard format … for an early warning notice?", as stated in the beginning to the response, the answer is no, but the contract requires (Clause 13.1) that all instructions, notifications, submissions, etc. are in a form that can be read, copied and recorded, so early warnings should not be a verbal communication such as a telephone conversation.

If the first notification is a telephone conversation, or a comment in a site meeting, which are quite likely, it should be immediately confirmed

Contract: .... Contract No:	EARLY WARNING NOTICE EWN No: ....

Section A: Enquiry

To: Project Manager/Contractor

Description

This matter could:

- ☐ Increase the total of the Prices
- ☐ Delay Completion
- ☐ Delay meeting a Key Date
- ☐ Impair the performance of the works in use

Risk Reduction meeting called? Yes/No Date: ....

Signed: .... (Contractor/Project Manager) Date: ....

Action by: Date required:

Section B: Reply

To: Contractor/Project Manager

Signed: .... (Contractor/Project Manager) Date: ....

Copied to:

Contractor ☐ Project Manager ☐ Supervisor ☐ File ☐ Other .... ☐

Figure 1.1 Suggested template for early warning notice

in writing in the format required by the contract to give it contractual significance.

Also, Clause 13.7 requires that notifications which the contract requires must be communicated separately from other communications; therefore, early warnings must not be included within a long letter which covers a numbers of issues or embodied within the minutes of a progress meeting.

There are some key words within the obligation to notify "The Contractor and the Project Manager". No one else has the authority or obligation to give an early warning. The Project Manager is therefore notifying on behalf of himself, the Employer, the Supervisor, the Employer's Designers and many possible others whom he represents within the contract.

The Contractor is notifying on behalf of himself, his Subcontractors, his Designers (if appropriate), and many possible others whom he represents under the contract. Early warnings should be notified by the key people named in Contract Data Part 2.

Project Managers are often criticised for seeing early warnings as something the Contractor has to do, but not the Project Manager, and in fact most early warnings are actually issued by the Contractor. However, the Contractor and the Project Manager are obliged to give early warnings each to the other, so it is critical that Project Managers play their part in the process and issue early warning notices where required.

As an example, if the Project Manager becomes aware that he will be late in delivering some design information to the Contractor because of a delay by a designer, he should issue the early warning as soon as he becomes aware that the information will not be delivered to the Contractor, not wait and subsequently blame the Contractor for not giving an early warning stating that he has not received the information!

- "As soon as" means *immediately*. A number of clauses within the contract deal with the situation in which the Contractor did not give an early warning. Whilst the party who gives the early warning must do so as soon as he becomes aware of the potential risk, the other party should respond as soon as possible and in all cases within the period for reply in Contract Data Part 1.
- "Could" does not mean *must*, *will* or *shall*. Clearly there is an obligation to notify even if it is only felt something may affect the contract, but there is no clear evidence that it will.

Example

On 1 February, the Contractor becomes aware of an issue which could increase the total of the Prices and delay Completion. He should immediately give an early warning to the Project Manager. However, he fails to do so until 22 February, three weeks later.

When the Contractor gives the early warning, the Project Manager notifies a compensation event and instructs the Contractor to submit a quotation, at the same time notifying the Contractor that he did not give an early warning of the matter that an experienced Contractor could have given.

If the Project Manager has given such notification, then the compensation event is assessed as if the Contractor had given early warning, so the Contractor should price his quotation based on the event as at 1 February, rather than 22 February.

If the Project Manager fails to give an early warning, there is no direct remedy within the contract, but the Project Manager would be losing an opportunity to raise an issue with the Contractor that could potentially present a risk to the success of the project, and if necessary to have a risk reduction meeting, so this failure could be detrimental to the Employer.

The Project Manager enters early warning matters in the Risk Register. Early warning of a matter for which a compensation event has previously been notified is not required. If the Project Manager gives an instruction for which a compensation event has already been notified, there is no requirement for either party to give an early warning.

It must be emphasised that early warnings are not the first step towards a compensation event as is often believed. Early warnings feature in a completely separate section of the contract and in fact the early warning provision is intended to prevent a compensation event occurring or at least to lessen its effect. It can also be used to notify a problem which is totally the risk of the notifier. It is also worth mentioning that early warnings are a notice of a future risk, not a past one.

The parties are not required, nor is it of any value, to notify a risk that has already happened.

Remedy for failure to give early warning

Under Clauses 61.5 and 63.5, if the Project Manager decides that the Contractor did not give an early warning of the event which an experienced contractor could have given, he notifies this to the Contractor when he instructs him to submit quotations. If the Project Manager has done so, the event is assessed as if the Contractor had given early warning.

Failure to give an early warning can also be considered as Disallowed Cost under Options C, D and E, Disallowed Cost being the cost which the Project Manager decides was incurred only because the Contractor did not give an early warning which the contract required him to give.

Question 1.2 What is a Risk Register and what is its purpose within the NEC3 Engineering and Construction Contract?

The Risk Register was first included within NEC3 in 2005 where it is defined under Clause 11.2(14) as "a register of the risks which are listed in the Contract Data and the risks which the Project Manager or the Contractor has notified as an early warning matter. It includes a description of the risk and a description of the actions which are to be taken to avoid or reduce the risk".

It is critical that the parties fully understand that the purpose of a Risk Register is to list all the identified risks and the results of their analysis and evaluation. It can then be used to track, review and monitor risks as they arise to enable the successful completion of the project.

The Risk Register does not allocate risk, as that is done by the contract.

In that sense, the contract does not prescribe the format or layout of the Risk Register, or its intended purpose, other than to list the risks in the contract and those that come to light at a later date and are notified as early warnings following which, if there is a risk reduction meeting, the Project Manager revises the Risk Register to record the outcome of the meeting.

Note that by the definition with Clause 11.2(4), risks which were not originally included in the Contract Data or subsequently notified as early warnings should not be included in the Risk Register.

Contract Data Part 2 allows the Contractor to identify matters which will be included in the Risk Register. The Risk Register does not allocate or change the risks in the contract, it records them and assists the parties in managing them. In that sense it is a valuable addition to the contract which was absent from previous editions but introduced in NEC3.

RISK REGISTER

Contract: ....

Contract No: ....

Contractor: ....
Project Manager: ....

Description of risk	Implications	Likelihood of occurrence (1–5) (Least – Most)	Potential impact (1–5) (Low – High)	Risk score	Risk owner	Mitigation strategy By whom? By when?	Allowance in the total of the Prices	Programme Allowance	Employer cost allowance	Risk status	Last updated
Discovery of unforeseen existing silo bases	Delay and additional costs in removal	2	4	8	Initially Contractor, unless additional silo bases discovered, then Employer	Site Information identifies likely locations; Contractor to take due regard	No allowance for unforeseen	No allowance for unforeseen	£20,000	Reducing Excavation 50% complete No additional silo bases discovered to date	15/12/2015

Figure 1.2 Typical Risk Register

A typical Risk Register which would comply with the contract would normally include the basic requirement for a description of the risk and a description of the actions which are to be taken to avoid or reduce the risk (see Figure 1.2).

There is no stated list of components of a Risk Register, but column headings should typically be titled as follows.

1 **Description of risk:** A clear description of the nature of the risk, if necessary referring to other documents such as site investigations, etc.
2 **Implications:** What would happen if the risk were to occur?
3 **Likelihood of occurrence:** This provides an assessment of how likely the risk is to occur. The example shows a forecast on a 1 (least likely) to 5 (most likely) basis, though it may be assessed as percentages, colour coding or simply "Low" (less than 30 per cent likelihood), "Medium" (31–70 per cent likelihood), or "High" (more than 70 per cent likelihood).
4 **Potential impact:** This assesses the impact that the occurrence of this risk would have on the project in terms of time and/or cost. The example shows the assessment on a 1 (low) to 5 (high) basis.

Risk management is essential to the success of any project and normally follows five steps:

Step 1 – Identify the risks
Step 2 – Decide who could be harmed and how
Step 3 – Evaluate the risks and decide on precautions
Step 4 – Record the findings and implement them
Step 5 – Review the assessment and update if necessary.

In order to compile the Risk Register the risks must first be listed, then they are quantified in terms of their likelihood of occurrence and their potential impact upon the project.

Question 1.3 What is the purpose of a risk reduction meeting within the NEC3 contracts?

In previous versions of NEC prior to NEC3, the "risk reduction meeting" was termed an "early warning meeting". Practitioners still state that would be the better term as it appears to be tied into the early warning process. However, the new name is better suited to what the meeting is for: it is aimed at reducing the risk that has been highlighted by the issue of an early warning notice by either Contractor or the Project Manager.

Under Clause 16.2, either the Project Manager or the Contractor may instruct the other to attend a risk reduction meeting.

The key word in Clause 16.2 is "instruct". If the Contractor calls a risk reduction meeting, he is instructing, not merely requesting, the Project Manager to attend. This clause is clearly intended to promote ownership of the project by the Contractor and the Project Manager, and any issues that could affect it, together with their resolution.

A risk reduction meeting need only have the Project Manager and the Contractor present as a minimum requirement, though each may instruct other people to attend if the other agrees, so in reality a number of people normally attend the meeting.

The "rule of meetings" will often apply in that the productivity of the meeting is often inversely proportional to the number of people who attend it!

Clause 16.3 describes what the attendees at a risk reduction meeting should do in an effort to either resolve the problem or at least attempt to resolve it:

- making and considering proposals for how the effect of the registered risks can be avoided or reduced ("registered" refers to the Risk Register)
- seeking solutions that will bring advantage to all those who will be affected
- deciding on the actions which will be taken and who, in accordance with this contract, will take them, and
- deciding which risks have now been avoided or have passed and can be removed from the Risk Register.

Clearly, from the above, the purpose of the risk reduction meeting is to actively consider ways to avoid or reduce the effect of the matter which has been notified. In some cases, the matter can be fully resolved, but in others, as the matter may not yet have occurred, it may simply have to be "parked" and recorded as such in the Risk Register. It is the Project Manager's responsibility to revise the Risk Register to record the outcome of the meeting, and to issue the revised Risk Register to the Contractor.

Chapter 2

Contractor's design

Submitting design proposals, liability for design, etc.

Question 2.1 We wish to use the NEC3 Engineering and Construction Contract and the Engineering and Construction Short Contract for a number of design and build projects. Can we do this and, if we can, how?

The NEC3 contracts, and in this case specifically the Engineering and Construction Contract and the Engineering and Construction Short Contract, are unique amongst construction contract families in that, whilst most other contracts provide for portions, or all of the design, to be carried out by the Contractor, they all use a separate contract for design and build where the Contractor has full responsibility for design, whereas the NEC3 contracts use the same contract whether design is to be carried out by Employer, the Contractor or a combination of the two, as responsibility for design, and any associated design criteria, performance requirements and obligations are defined within the Works Information.

In reality, there is no need for a separate design and build contract as within the NEC3 contracts the clauses which cover early warnings, programme, payment, change management, etc. are the same whether or not the Contractor has design responsibility.

If we consider the NEC3 Engineering and Construction Contract, the "default position" within the contract is that all design is carried out by the Employer, but any design to be carried out by the Contractor is assigned to him within the Works Information, and Clauses 21 to 23 cover design obligations, and submission and acceptance of the Contractor's design proposals.

In the Engineering and Construction Contract "the Contractor provides the Works in accordance with the Works Information" (Clause 20.1) and "the Contractor designs the parts of the works which the Works Information states he is to design" (Clause 21.1).

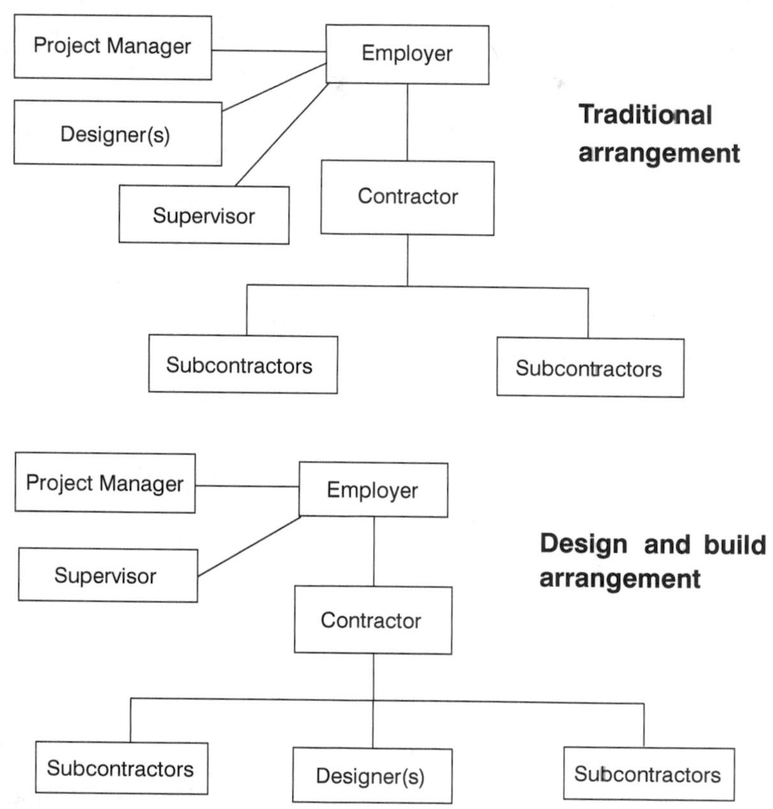

Figure 2.1 Design and Build using the NEC3 Engineering and Construction Contract

Any design to be carried out by the Employer within the Engineering and Construction Short Contract is also assigned within the Works Information in the same way.

It is clear, then, that the Works Information should clearly show what, if anything, is to be designed by the Contractor and this could consist of detailed specifications or some form of performance criteria and certain warranties.

When preparing the Works Information, which includes Contractor's design, it is important that the right balance of information should be considered. If it is too prescriptive it will lead to all the tendering Contractors submitting similar designs and prices. If the Works Information is not detailed enough, then none of the tendering Contractors will produce a design which is as the Employer intended.

Question 2.2 In a design and build contract using the NEC3 Engineering and Construction Contract, which takes precedence: the Employer's Requirements or the Contractor's Proposals?

The NEC3 contracts are unlike most other construction contracts in that they do not provide a priority or hierarchy of documents, but one can "read into" the various clauses to determine which takes precedence in the event of an inconsistency.

A question to then be considered is: If the Contractor has full design responsibility and there is an inconsistency between the Works Information provided by the Employer (normally referred to in other contracts as the "Employer's Requirements") and the Works Information provided by the Contractor (normally referred to in other contracts as the "Contractor's Proposals"), which takes precedence?

The answer can be found in Clauses 11.2(5) and 60.1(1).

Defects are defined in Clause 11.2(5) as:

- a part of the works which is not in accordance with the Works Information, or
- a part of the works designed by the Contractor which is not in accordance with the applicable law or the Contractor's design which the Project Manager has accepted.

From bullet point 1 above, if the Contractor's design does not comply with the Works Information, then it is a Defect, the Works Information clearly taking precedence over the Contractor's design.

From bullet point 2, the Contractor's Design must also comply with the law and anything that the Project Manager has already accepted.

Under Clause 60.1(1), the Project Manager gives an instruction changing the Works Information except for:

- a change made in order to accept a Defect, or
- a change to the Works Information provided by the Contractor for his design which is made either at his request or to comply with other Works Information provided by the Employer.

From bullet point 2 above, if the Contractor has to change his design to comply with the Works Information provided by the Employer, that is not a compensation event, again reinforcing the fact that the Works Information provided by the Employer takes precedence.

Question 2.3 If the Project Manager on an NEC3 Engineering and Construction contract accepts the Contractor's design and it is later found that the design will not work, or it is not approved by a third party regulator, who is liable?

Under Clause 21.2, the Contractor is required to submit the particulars of his design as the Works Information requires to the Project Manager for acceptance.

The particulars of the design in terms of drawings, specification and other details must clearly be sufficient for the Project Manager to make the decision as to whether the particulars comply with the Works Information and also, if relevant, the applicable law. The Works Information may stipulate whether the design may be submitted in parts, and also how long the Project Manager requires to accept the design.

Note that in the absence of a stated timescale for acceptance/non-acceptance of design, the "period for reply" will apply.

Many Project Managers are concerned that they have to "approve" the Contractor's design and therefore they are concerned as to whether they would be qualified, experienced and also insured to be able to do so. Note the use of the word "acceptance" as distinct from "approval". Acceptance denotes compliance with the Works Information or the applicable law. It does not denote that the design will work, that it will be approved by regulating authorities, or that it will fulfil all the obligations which the contract and the law impose. Therefore, performance requirements such as structural strength, insulation qualities, etc., do not need to be considered.

A reason for the Project Manager not accepting the Contractor's design is that it does not comply with the Works Information or the applicable law. The Contractor cannot proceed with the relevant work until the Project Manager has accepted the design.

Note that under Clause 14.1, the Project Manager's acceptance does not change the Contractor's responsibility to Provide the Works or his liability for his design. This is an important aspect as, if the Project Manager accepts the Contractor's design but following acceptance there are problems with the proposed design – for example, in meeting the appropriate legislation or the requirements of external regulatory bodies – this is the Contractor's liability.

Note that, under Clause 27.1, the Contractor is required to obtain approval of his design from Others where necessary. This will include planning authorities, and other third party regulatory bodies.

Note also, Clause 60.1(1) second bullet point, which clearly refers to the fact that the Employer's Works Information prevails over the Contractor's Works Information. This clause gives precedence to the Works Information in Part 1 of the Contract Data over the Works Information in Part 2 of the Contract Data. Thus the Contractor should ensure that the Works Information he prepares and submits with his tender as Part 2 of the Contract Data, complies with the requirements of the Works Information in Part 1 of the Contract Data. 2.7.

Question 2.4 We wish to novate a design consultant from the Employer to the Contractor under an NEC3 Engineering and Construction Contract with full Contractor's design. Can we do this?

As stated previously, whilst the principle of design and build agreements is that the Employer preparers his requirements and sends them to the tendering Contractors and they prepare their proposals to match the Employer's requirements, in reality in nearly half of design and build contracts the Employer has already appointed a design team which prepare feasibility proposals and initial design proposals before tenders are invited.

Outline planning permission and sometimes detailed permission may have been obtained for the scheme before the Contractor is appointed. Each Contractor then tenders on the basis that the Employer's design team will be novated or transferred to the successful tendering Contractor who will then be responsible for appointing the team and completing the design under a new agreement. This process is often referred to as "novation", which means "replace" or "substitute", and is a mechanism where one party transfers all its obligations and benefits under a contract to a third party. The third party effectively replaces the original party as a party to that contract, so the Contractor is in the same position as if he had been the Employer from the commencement of the original contract.

Many prefer to use the term "consultant switch" where the design consultant "switches" to work for the Contractor under different terms as a more accurate definition. This approach allows the Employer and his advisers time to develop their thoughts and requirements, consider planning consent issues, then when the design is fairly well advanced, the designers can be passed to the successful design-and-build Contractor.

The NEC3 contracts do not include pro forma novation agreements, but it is critical that the wording of these agreements be carefully considered, as there are many badly drafted agreements in existence. Novation can be by a signed agreement or by deed. As with all contracts, there must be consideration, which is usually assumed to be the discharge of the original contract and the original parties' contractual obligations to each other. If the consideration is unclear, or where there is none, the novation agreement should be executed as a deed. In many cases the agreement states briefly (and very badly) that from the date of the execution of the novation agreement, the Contractor will take the place of the Employer as if he had employed the designers from the beginning, the document stating that in place of the word "Employer" one should read "Contractor". However, the issues of design liability, inspection, guarantees and warranties may not always apply on a back-to-back basis, so one must take care to draft the agreement in sufficient detail and refer to the correct parties.

Question 2.5 In an NEC3 Engineering and Construction Contract, when would we use Option X15 (limitation of the Contractor's liability for his design to reasonable skill and care)?

Whilst the Works Information defines what, if any, design is to be carried out by the Contractor, the contract is silent on the standard of care to be exercised by the Contractor when carrying out any design.

Two terms that relate to design liability are "fitness for purpose" and "reasonable skill and care".

Whilst this book is intended for international use, in defining the term "fitness for purpose" one must look in English law to the Sale of Goods Act 1979, which refers to the quality of goods supplied including their state and condition complying in terms of "fitness for all the purposes for which goods of the kind in question are commonly supplied".

In construction, fitness for purpose means producing a finished project fit in all respects for its intended purpose. This is an absolute duty independent of negligence, and in the absence of any express terms within the contract to the contrary, a Contractor who has a design responsibility will be required to design and build the project "fit for purpose".

Some contracts will limit the Contractor's liability to that of a consultant, i.e. reasonable skill and care. The Engineering and Construction Contract does this through Option X15. If this Secondary Option is not chosen, the Contractor's liability for design is "fitness for purpose".

Contrast this with the level of liability of a consultant (whether acting for an Employer or a Contractor) in providing a design service, which is defined, again in English law by the Supply of Goods and Services Act 1982, where there is an implied term that the consultants will carry out the service, in this case design, with reasonable care and skill, which means designing to the level of an ordinary, but competent person exercising a particular skill.

So, in the absence of any express terms to the contrary, a designer will normally be required to design using "reasonable skill and care". This is normally achieved by the designer following accepted practice and complying with national standards, codes of practice, etc.

Clearly, despite the Contractor believing that he has offset his design obligations and liability to his designing consultant, he has to be aware that he and his consultant have differing levels of care and liability.

Question 2.6 The Contractor has designed a temporary access walkway bridging across two areas of the Site. He has not submitted his design proposals for the Project Manager's acceptance before constructing the walkway, but we insist that he is in breach of contract by failing to do so. What can we do to remedy this breach?

If the Contractor has designed an item of Equipment, for example a temporary access way, or specialist scaffold, the Project Manager, under Clause 23.1 *may* instruct him to submit particulars to him for acceptance, so the submission is not obligatory as with the design of parts of the works.

So, in this case, unless the Project Manager gave that instruction, the Contractor is not in breach of contract by failing to submit particulars of his design.

Notwithstanding this, the Contractor's design, whether the Project Manager has given the instruction or not, the piece of work still has to be built so that it is safe and fully compliant with any relevant legislation.

The Project Manager may not accept the design if it does not comply with the Works Information, the Contractor's design which the Project Manager has accepted or the applicable law.

Chapter 3

Time and the Accepted Programme

The submission or non-submission of a programme, float and time risk allowances, method statements, etc.

Question 3.1 Can we include liquidated or unliquidated damages in an NEC3 contract? How are these deducted in the event of delayed Completion?

Liquidated damages in other construction contracts are referred to as delay damages in the NEC3 contracts, and in the NEC3 Engineering and Construction Contract are covered by Secondary Option X7.

Delay damages are pre-defined amounts inserted into the contract and paid or withheld from the Contractor in the event that he fails to complete the works by the Completion Date. The amount included within the contract for delay damages should be a "genuine pre-estimate of likely loss" – i.e. should not constitute a penalty.

Many contracts require the contract administrator (Engineer, Architect, etc.) to issue some form of certificate confirming that the Contractor failed to complete on time, and also for the Employer to notify the Contractor in writing that he will be withholding the relevant damages, but the Engineering and Construction Contract provides for the Contractor to pay delay damages at the rate stated in the Contract Data until Completion or the date on which the Employer has taken over the works, whichever is earlier, without having to certify that the Contractor has defaulted.

Also, many contracts require the contract administrator to value the works and the Employer then deducts the liquidated damages from the amount due to the Contractor, but the Engineering and Construction Contract requires the Project Manager to deduct amounts to be paid or retained from the Contractor in respect of delay damages within his assessment and certificate.

Under Clause X7.3, if the Employer takes over the works before Completion, the Project Manager assesses the benefit to the Employer of taking over that part of the works as a proportion of taking over the whole of the works and the delay damages are reduced in this proportion.

Whilst it is probably the most correct way to assess remaining delay damages, the Project Manager having to assess the benefit to the Employer can at best be a subjective exercise and may possibly lead to disputes with the Contractor.

Most other contracts state that the amount of liquidated damages is reduced by the same proportion the part that has been taken over bears to the value of the whole of the works, so if a third of the value of the works has been taken over, the amount of the liquidated damages is reduced by one-third.

Option X7 is included within:

- the Engineering and Construction Contract
- the Professional Services Contract.

If Secondary Option X7 is not selected, and the Contractor fails to complete the works by the Completion Date, he may be liable to the Employer for unliquidated damages, though in this case the Employer would have to prove and quantify any loss or cost incurred as a result of that failure, and also, of course, try to recover those damages from the Contractor as the contract does not allow the Project Manager to deduct amounts to be paid or retained from the Contractor, or for the Employer to withhold payment on this respect.

Question 3.2 Do the NEC3 contracts have provision for Sectional Completion where the Employer may wish to take over parts of the works?

If the Employer requires sections of the works to be completed by the Contractor before the whole of the works are completed, then Secondary Option X5 should be chosen.

References in the contract to the works, Completion and Completion Date will then apply to either the whole of the works, or a section.

Contract Data Part 1 should give a description of each section and the date by which it is to be completed. Option X5 may be selected together with Option X6 (bonus for early Completion) and/or Option X7 (Delay damages).

Option X5 is included within:

- the Engineering and Construction Contract
- the Professional Services Contract.

Question 3.3 In an NEC3 Engineering and Construction Contract, what is the relationship between "Completion" and "Take Over"?

In order to consider the relationship between these two terms, let us first consider their definitions under the Contract.

Completion: The Engineering and Construction Contract defines Completion under Clause 11.2(2) as when the Contractor has:

- done all that the Works Information requires him to do by the Completion Date, and
- corrected notified defects which would have prevented the Employer from using the works and Others from doing their work.

Take over: The Employer may use any part of the works before Completion has been certified. If he does so, he takes over the part of the works when he begins to use it except if the use is:

- for a reason stated in the Works Information, or
- to suit the Contractor's method of working.

The Project Manager certifies the date upon which the Employer takes over any part of the works and its extent within one week of that date.

Once the Contractor has completed the works, the Employer takes over within two weeks. This occurs even if the Contractor completes the works early. However, there is an optional statement within Contract Data Part 1 that "the Employer is not willing to take over the works before the Completion Date". If this is selected then if the Contractor completes the works early he still has responsibility for the works until the Completion Date.

Note that the defects date is stated in Contract Data Part 1 as 'X' weeks after Completion of the whole of the works.

If Option X7 is selected, the liability for delay damages is calculated from the Completion Date until the earlier of:

- Completion, and
- the date on which the Employer takes over the works.

Partial possession

Whilst many contracts have express provision for partial possession by the Employer, with the consent of the Contractor which shall not be unreasonably withheld, the Engineering and Construction Contract does not, though under Clause 35.2 the Employer may use any part of the works before Completion has been certified.

Take over of that part is deemed to have occurred unless otherwise stated in the Works Information, or if it has been done to suit the Contractor's method of working.

Note that, if the Project Manager certifies take over of a part of the works before Completion and the Completion Date, then it is a compensation event under Clause 60.1(15) if the Employer takes over the works before Completion but after the Completion Date then the Contractor is already late in completing and therefore that would not be a compensation event.

There is no provision for the Contractor to give or refuse consent to the Employer taking over the works. The Project Manager certifies the date of take over, or partial take over within one week of it taking place.

Question 3.4 We have a project where we require the Contractor to provide "as built" drawings, maintenance manuals and staff training as part of the contract. How do we include this requirement within the contract?

Once the project is built, it is often very difficult for Employers to obtain supporting documents such as "as built" drawings, maintenance manuals, software licences, etc. In order to consider and remedy this issue, let us first consider how the NEC3 contracts consider "completion".

Whilst many contracts use the terms "Practical Completion" and "Substantial Completion", the NEC3 contracts do not, as these terms are often subject to various interpretations including whether the Employer can take beneficial occupation of the "completed" works.

The Engineering and Construction Contract, for example, defines Completion under Clause 11.2(2) as when the Contractor has:

- done all that the Works Information requires him to do by the Completion Date, and
- corrected notified defects which would have prevented the Employer from using the works and Others from doing their work.

The first bullet requires the Contractor to comply with the requirements of the Works Information and supersedes traditional term such as "practical completion". This requirement may include, not only physical completion of the Works, but also submission of "as built" drawings, maintenance manuals, training requirements, successful testing requirements, etc.

By including these requirements within the Works Information, completion has not been achieved until they are completed. If Option X7 (delay damages) has been selected, the Contractor would also be liable for delay damages until the earlier of:

- Completion and
- the date on which the Employer takes over the works.

Referring to the second bullet, the works may contain defects, though these are not significant enough to prevent the Employer from practically and safely using the works. Completion is certified with the requirement that the Contractor corrects the defect before the end of the defect correction period following Completion.

There is also a fall back within Clause 11.2(2) in that if the work which the Contractor is to do by the Completion Date is not stated in the Works Information, Completion is when the Contractor has done all the work necessary for the Employer to use the works and for Others to do their work. This broadly aligns with the traditional terms "Practical Completion" and "Substantial Completion", where the Employer is able to take beneficial occupation of the works and use them as intended.

The Project Manager is responsible for certifying Completion, as defined in Clause 11.2(2), within one week of Completion. Normally, the Contractor will request the certificate as soon as he considers he is entitled to it, but such a request is not essential.

The Engineering and Construction Short contract uses similar wording to the Engineering and Construction Contract, with the Professional Services Contract referring to the Consultant having done all which the "Scope" states he is to do by the Completion Date, instead of the "Works Information".

Question 3.5 We have a project under the NEC3 Engineering and Construction Contract Option C (target contract with activity schedule) to build a new college where we will require the Contractor to complete wall finishings to classrooms in order that a third party directly employed by the Employer will be able to install audio visual equipment fixed to the walls. How can we incorporate this requirement into the contract?

First, let us just pick up on some terminology within the NEC3 contracts. "Equipment" is defined under Clause 11.2(7) as "items provided by the Contractor and used by him to Provide the Works and which the Works Information does not require him to include in the works". So, should the audio visual equipment be referred to as either Plant or Materials which are defined under Clause 11.2(12) as "items intended to be included in the works"?

The requirement for the Contractor to complete certain works can be included within an NEC3 Engineering and Construction Contract by using Key Dates.

Example

The Contractor is building a new outpatients department for a local hospital. The Employer wishes to have new X-ray machinery installed by a specialist whom he will employ directly and who will install the machinery as the building work progresses.

In order to do that the Contractor has to complete the part of the building that will house the machinery and install the necessary electrical power facilities so that the machinery can be tested prior to completion of the project. The necessary work to be carried out by the Contractor in readiness for the specialist to carry out his part will be the Condition, and this will have a Key Date attached to it.

If the Contractor then fails to meet the Condition by the Key Date and the Employer incurs a cost in having to postpone the installation of the X-ray machinery, then this cost is paid by the Contractor. The cost must be incurred on the same project. So, for example, if the late installation of the X-ray machinery incurs a cost on another project, then this is not paid by the Contractor.

Within the Engineering and Construction Contract this is covered by Clause 11.2(9), a Key Date being defined as "the date by which work has to meet the Condition stated".

It is an optional clause allowing the Employer, should he wish, to include dates when the Contractor must complete certain items of work, possibly to allow others to carry out other work. If used the Condition and the applicable Key Date are defined by the Employer in Contract Data Part 1.

"Key Dates" must be differentiated from "Completion" or "Sectional Completion", in that with Key Dates the Employer does not take over the works, they stay under the possession and control of the Contractor, whereas with Completion and Sectional Completion, the Employer takes over the works. In this respect there are no pre-determined delay damages applied to the Contractor who does not meet a Condition by a Key Date.

However, if the Project Manager decides that the Contractor has not met the Condition stated by the Key Date, and the Employer incurs additional cost on the same project as a result of that failure, then the Contractor will be liable to pay that amount (Clause 25.3). By referring to "the same project", the Employer cannot claim costs incurred on another project as a result of the failure of the Contractor to meet a Key Date. This cost is assessed by the Project Manager within four weeks of when the Contractor actually does meet the Condition for the Key Date.

Key Dates are provided in NEC3 contracts as follows:

- Engineering and Construction Contract
- Professional Services Contract.

Question 3.6 The Contractor on an Engineering and Construction Contract was required to submit a programme with his tender. He has stated that as we accepted his tender, then we have also accepted his programme, and this is therefore the first Accepted Programme, but how can it be so if it does not comply with Clause 31.2?

Employers should instruct Contractors to submit at least outline programmes with their tenders, though it must be appreciated that as the Contractor may not be successful with his tender it will serve only as an indicative outline of how the Contractor would carry out the works should he be successful.

The Engineering and Construction Contract provides as an optional statement for the Contractor to submit a first programme for acceptance within a specified number of weeks of the Contract Date.

So the two options for the Contractor's submission of his first programme for acceptance are as follows:

1 He may submit, or be required to submit, a programme with his tender, in which case it is referenced by the Contractor in Contract Data Part 2. (Note that under Clause 11.1(1) "the Accepted Programme is the programme identified in the Contract Data or is the latest programme accepted by the Project Manager". Therefore, any programme referred to in Contract Data Part 2 automatically becomes the first Accepted Programme even though it may not necessarily comply fully with the requirements of Clause 31.2.)
2 If a programme is not identified in Contract Data Part 2, the Contractor submits a first programme to the Project Manager for acceptance within the period of time after the Contract Date, this period of time being stated in Contract Data Part 1.

Question 3.7 Who owns the float in an Accepted Programme within an NEC3 Engineering and Construction Contract?

Float is any spare time within the Contractor's programme, after time risk allowances have been included, and represents the amount of time that operations may be delayed without delaying following operations and/or planned Completion.

It can also represent the time between when the Contractor plans to complete and when the contract requires him to complete (terminal float). Float absorbs to a certain extent the Contractor's own delays or the delays caused by a compensation event, thereby lessening or avoiding any delay to planned Completion. In effect, no delay arises unless float on the relevant and critical operations reduces to below zero. Programming is never an exact science, so float gives some flexibility to the Contractor in respect of incorrect forecasts or his own inefficiencies. As the work progresses the float will change as output rates change, Contractor's risk events take place and also compensation events arise.

The general belief, certainly amongst many Contractors, is that float belongs to them as they wrote the programme, and therefore they have the right to work to that programme and use any float that it contains.

Example

Completion Date is 20 November 2015, but the Contractor is planning to complete by 6 November 2015, the programme showing the last operation (roof covering) being completed on that date.

The Project Manager has accepted the Contractor's programme. At the beginning of October 2015, the Project Manager gives an instruction to change the specification for the roof from concrete tiles to slates. This is a change to the Works Information and therefore a compensation event under Clause 60.1(1).

The Contractor prepares his quotation and finds that he cannot get the slates delivered until week commencing 9 November 2015. Therefore, the roof cannot be completed until 13 November 2015, and the Contractor then shows in his quotation a delayed to planned Completion of one week.

As Clause 63.3 states that "a delay to the Completion Date is assessed as the length of time that, due to a compensation event, planned Completion is later than planned Completion as shown on the Accepted Programme", assuming the Project Manager accepts the quotation, planned Completion becomes 13 November 2015 and the Completion Date becomes 27 November 2015.

Conversely, Employers believe that if it is free time then they have the right to use it, but it actually depends where the float is and what it is for. In principle, float other than terminal float or time risk allowances, is a shared resource – it is spare time, it belongs to the project, and therefore may be used by whichever party needs it first.

There are three primary types of float:

1 the amount of time that an operation can be delayed before it delays the earliest start of following operation (free float)
2 the amount of time that an operation can be delayed before it delays the earliest completion of the works (total float)
3 the amount of time between planned Completion and the Completion Date (terminal float).

Generally, with other contracts, if the Contractor shows that he plans to complete a project early, then he is prevented from completing as early as he planned, but he still completes before the contract completion date, then although there may be an entitlement to an extension of time under the contract – for example, exceptionally adverse weather – none will be awarded as there is no delay to contract completion, but he may have a right to financial recovery subject to him proving loss and/or expense. The Engineering and Construction Contract deals with this in a different way.

If the Contractor shows on his programme planned Completion earlier than the Completion Date and he is prevented from completing by the planned Completion Date by a compensation event, then, when assessing the compensation event, Clause 63.3 states "a delay to the Completion Date is assessed as the length of time that, due to a compensation event, planned Completion is later than planned Completion as shown on the Accepted Programme", therefore any terminal float is retained by the Contractor, the period of delay being added to the Completion Date to determine the change to the Completion Date.

Any delay to planned Completion due to a compensation event thus results in the same delay to the Completion Date. Therefore, the extension is granted on the basis of time the Contractor is delayed, i.e. entitlement, not on how long he needs to achieve the current Completion Date.

Question 3.8 How often must the Contractor in an NEC3 Engineering and Construction Contract submit a revised programme?

The timing for submission of revised programmes for acceptance is defined by Clause 32.2:

- within the period for reply after the Project Manager has instructed him (for example, the Project Manager could instruct the Contractor to submit a programme to enable an issue to be discussed in a risk reduction meeting)
- when the Contractor chooses (the Contractor may feel it beneficial to submit a programme to be discussed in a risk reduction meeting), and
- in any case at intervals no longer than that stated in Contract Data Part 1 (this is the longest period for submission of a revised programme).

The revised programme shows the following:

- The actual progress achieved on each operation and its effect upon timing of the remaining work The progress achieved may be shown on the programme itself or appended to it in the form of a progress report. Operations not yet completed will normally be shown as percentage complete, being the progress achieved to date against the total work within the operation on the date the assessment is made. It is useful if the Contractor and the Project Manager agree the progress statement before the revised programme comes into being. The submitted revised programme is then an undisputed statement of fact.
- The revised programme should reflect "planned vs actual" progress including early or delayed start, early or delayed completion of each operation. It is important that, if the Contractor states a percentage completed to date, he should show actual progress achieved rather than time expired to date. Also, the Contractor should not just make statements such as "operation on programme" as this may be misleading.
- An operation on programme does not necessarily mean that the operation is not delayed, for example the Contractor is expecting to complete an operation late, then stating the words "operation on programme" can be interpreted as either in accordance with the contract, or progressing late in accordance with his expectations.
- It is also important to recognise that many operations may not progress on a "straight line" basis, for example building a brick wall may initially involve work and resources in setting out and building corners, with the bulk areas following on.
- The effects of implemented compensation events.

The June 2006 amendment to the contract changed the sub-clause from "the effects of implemented compensation events and of notified early warning matters" to "the effects of implemented compensation events". It is perhaps curious that the effects of notified early warning matters are not required to be shown on revised programmes.

Note that a compensation event is not "implemented" until the quotation has been accepted or assessed by the Project Manager.

Whilst the completion date is not changed until a compensation event has been implemented, there is a danger in only showing compensation events that have been implemented as the revised programme is not reflecting reality.

- How the Contractor plans to deal with any delays and to correct notified Defects. The Contractor should show in his revised programme any delays which may or may not be caused by him, and also time needed to correct his own Defects.

- Any other changes which the Contractor proposes to make to the Accepted Programme.
- An example of this could be that the Contractor resequences part of the work or proposes to use a different method or different Equipment to that specified by him in a previous method statement.

It is therefore very much a "living" programme. If the Contractor does not submit a programme which the contract requires, then the Project Manager can assess compensation events, without receiving a quotation from the Contractor (Clauses 64.1 and 64.2).

The Accepted Programme:

- effectively provides an agreed record of the progress of the job and where the delays have come from
- provides a realistic base for future planning by both the Contractor and Project Manager
- is the base from which changes to the Completion Date are calculated
- is the base from which additional costs are derived (because each operation has a method statement and resources attached to it, the change in resources can be calculated and hence the change in costs; because information is so much more transparent, there is more scope for working together).

Example

The Project Manager instructs the Contractor not to paint the walls to certain rooms of a new building, as he is considering an alternative finish which he will instruct at a later date. The following week, the Contractor is due to submit a revised programme, but he has only just submitted his quotation to the Project Manager for the omission of the painting, so there is probably a couple of weeks before the Project Manager replies and the compensation event is implemented. Should the Contractor in the meantime be showing the walls as to be painted or not to be painted, as the compensation event has not yet been implemented? It should show the walls as not be painted as it reflects reality.

Question 3.9 In an NEC3 Engineering and Construction Contract being carried out under Option A (priced contract with activity schedule), can we require the Contractor to bring forward Completion if the Employer has a need to take over a project earlier?

Acceleration in many contracts normally means increasing resources, working faster or working longer hours, so that completion can be achieved *by* the completion date in the contract. In effect the Contractor is catching up to recover a delay.

However, within the NEC3 contracts, acceleration means bringing increasing resources, working faster or working longer hours so that Completion can be achieved *before* the Completion Date.

Under Clause 36.1 of the Engineering and Construction Contract, the Project Manager may instruct the Contractor to submit a quotation for acceleration. He also states any changes to Key Dates to be included within the quotation. The quotation must include proposed changes to the Prices, and a revised programme showing the earlier Completion Date and changed Key Dates.

As with quotations for compensation events, the Contractor must submit details within his quotation; however, the contract does not prescribe how the quotation is to be priced – for example, based on Schedule of Cost Components.

There is also no remedy if the Contractor does not submit a quotation, though under Clause 36.2, the Contractor submits his quotation or gives his reasons for not doing so within the "period for reply".

Assessment of a quotation for acceleration by the Project Manager is different from that of a compensation event in that, if the Contractor's quotation for a compensation event is not accepted by the Project Manager, he can make his own assessment.

Acceleration can only be undertaken by agreement between the Project Manager and the Contractor and cannot be imposed on the Contractor or any assessment imposed upon him without his agreement. When the Project Manager has accepted the quotation for acceleration, he changes the Prices, Completion Date and Key Dates and accepts the revised programme.

Acceleration is a provision within the:

- Engineering and Construction Contract
- Engineering and Construction Subcontract
- Professional Services Contract.

Chapter 4

Testing and Defects

The roles of the Parties, testing requirements, searching for Defects, etc.

Question 4.1 We have identified a number of Defects on a project which a Contractor is carrying out under the NEC3 Engineering and Construction Contract, but the Contractor denies that they are Defects. How do the NEC3 contracts define a Defect?

A Defect is defined within the contract (Clause 11.2(5)) as:

- a part of the works which is not in accordance with the Works Information, or
- a part of the works designed by the Contractor which is not in accordance with the applicable law or the Contractor's design which the Project Manager has accepted.

In the first bullet, the Works Information provides the reference point for what the Contractor has to do to Provide the Works. If the work done by the Contractor does not comply with the Works Information, then unless the Defect is accepted (Clause 44), the Contractor is obliged to correct the Defect.

In most cases the work done will fall short of the Works Information, but as the words "not in accordance with" are used, it is a compliance issue. Therefore, work which exceeds the requirement within the Works Information would also be a Defect. Clearly, the Project Manager may wish, having discussed with Employer, to accept such a Defect!

A Defect may not necessarily mean that the work is not fit for purpose in that the Defect could simply be a colour variation – for example, the Works Information stated a certain shade of red paint and the Contractor used a different shade of red paint – in which case a Defect exists as that part of the works was not in accordance with the

Works Information. In such a case, unless the shade of red is a particular corporate colour or a stipulation of a regulatory authority it may be prudent to accept the Defect.

It is also possible that work is "defective" but as it complies with the Works Information, it is not a Defect! For example, the specification within the Works Information may stipulate use of a particular material which then fails. In that respect the work is defective but as it complies with the Works Information, it is actually not a Defect.

In the second bullet, if the Project Manager accepts the Contractor's design under Clause 21.2, then subsequently the Contractor either does not comply with the applicable law, or changes the design, then again that is a Defect. If the Contractor wishes to change his design then he has to re-submit the new design to the Project Manager for his acceptance. (See Figure 4.1.)

Until the defects date, the Supervisor is obliged to notify the Contractor of every Defect as soon as he finds it, and the Contractor is obliged to notify the Supervisor as soon as he finds it.

Whilst the requirement for the Supervisor to notify the Contractor of each Defect is useful as the Contractor may not have been aware of the Defect or that it was a Defect, so he can correct it immediately, the obligation for the Contractor to notify the Supervisor of each Defect as soon as he finds it is somewhat cumbersome as the Supervisor probably does not want or need to know about every Defect, particularly if the Contractor is already correcting it.

Question 4.2 On an Engineering and Construction Contract using Option C (target contract with activity schedule), the Contractor has advised us that he is entitled to be paid for correcting a Defect; we disagree that we should be paying him for correcting Defects. Who is correct?

The Contractor is liable for correction of Defects and, as a general rule, he is not entitled to be paid additional monies or given additional time in which to correct those Defects.

However, note that under Options C, D and E where the Contractor is reimbursed his Defined Costs, two Disallowed Costs under Clause 11.2(25) deal specifically with Defects.

The first bullet point states the cost of:

- correcting Defects after Completion.

Contract:	DEFECTS NOTIFICATION
Contract No:	D/N No: ....

The Contractor is notified of the following defects:

Description	Date corrected

The defect correction period is:

Certified all defects corrected

.................................(Supervisor) Date: ....

Copied to:

Contractor ☐ Project Manager ☐ Supervisor ☐ File ☐ Other .... ☐

Figure 4.1 Defects notice

If the Contractor corrects Defects *after* Completion then he *is not* reimbursed the Defined Cost of doing so. However, and this is where Employers and many NEC3 practitioners express some concern, if the Contractor corrects Defects *before* Completion then he *is* reimbursed the Defined Cost of doing so.

The threshold is when the Defect is corrected, not when it occurs, so a Defect which appears before Completion, but is corrected after Completion is a Disallowed Cost.

There is no limit to the type or size of Defects, or why the Defect occurred, e.g. through the carelessness of the Contractor.

Whilst some may see this as the Contractor's right to repeatedly attempt to get the work right at the Employer's expense, it must be remembered that in the case of Options C and D these are target contracts, therefore if the Contractor is paid for correcting a Defect, not only is this bad for his reputation, but as it is not a compensation event he is not being given additional time to correct the Defect and the additional cost paid will reduce his entitlement to the Contractor's share at Completion.

The second bullet point states the cost of:

- correcting Defects caused by the Contractor not complying with a constraint on how he is to Provide the Works stated in the Works Information.

So, if the Contractor has to correct a Defect because he did not comply with the Works Information, whether it is before or after Completion, then he will not be reimbursed the cost of doing so.

Question 4.3 What is the meaning of the term "search for a Defect" under Clause 42.1 of the NEC3 Engineering and Construction Contract?

Under Clause 42.1, the Supervisor (not the Project Manager, unless the Supervisor has delegated the authority to him) has the authority to instruct the Contractor to search. This action is normally defined as "uncovering", "opening up" or "uncovering" in other contracts. It is usually required in order to investigate whether a defect exists, and possibly the cause, and the corrective measures required. Searching can include uncovering, dismantling, reassembly, providing materials and samples and additional tests which the Works Information did not originally require.

If the Supervisor instructs the Contractor to search for a Defect and no Defect is found, this is a compensation event (Clause 60.1(10)). However, if the search is needed only because the Contractor gave insufficient notice of doing work obstructing a required test or inspection, then it is not a compensation event.

Question 4.4 What can be done on an NEC3 Engineering and Construction Contract if the Contractor fails to correct Defects?

The Project Manager arranges for the Employer to allow access to parts of the works taken over in order to correct a defect.

If the Contractor does not correct the Defect within the defect correction period in the contract, the Employer assesses the cost of having the Defect corrected by other people and the Contractor pays this amount.

Note that the Employer may choose not to have the Defect corrected by anyone, but that is his prerogative, the Contractor still pays this amount.

Question 4.5 When is the Defects Certificate issued under an NEC3 Engineering and Construction Contract, and who issues it?

The Defects Certificate is issued by the Supervisor to the Project Manager and the Contractor at the later of the defects date and the end of the last defect correction period (see Figure 4.2).

Some practitioners state that as the Project Manager issues the Completion Certificate, he should also issue the Defects Certificate, which is common with most other contracts, but by requiring the Supervisor to issue the Defects Certificate, the Engineering and Construction Contract reinforces the role and responsibility of the Supervisor in dealing with Defects.

Whilst many contracts require any defects to be corrected before the Defects Certificate, or its equivalent is issued, under the Engineering and Construction Contract the Defects Certificate is issued by the Supervisor at the appropriate time and may either list Defects which the Contractor has not corrected, or a statement that there are no outstanding defects.

Note that if Option X16 (Retention) has been selected, under Clause X16.2, the amount retained is halved:

- in the assessment made at Completion of the whole of the works, or
- in the next assessment after the Employer has taken over the whole of the works if this is before completion of the whole of the works.

The amount retained remains at this amount until the Defects Certificate is issued. No amount is retained in the assessments made after the Defects Certificate has been issued.

Note that this may mean that the retention is released to the Contractor, even though there are still outstanding Defects yet to be corrected.

The Defects Certificate is not conclusive in that if a Defects Certificate states that there are no defects it does not prevent the Employer from exercising his rights should a defect arise later or the Supervisor did not find or notify it.

However, the Employer may wish to take action against a Supervisor who did not carry out the level of inspection he had paid him to do! The Supervisor will normally only list what are usually defined as "patent defects", i.e. those which are observable from reasonable inspection at the time, examples being a defective concrete finish or an incorrect paint colour and may not include what are usually defined as "latent defects", which may be hidden from reasonable inspection and may come to light at a later date, examples being some structural defects.

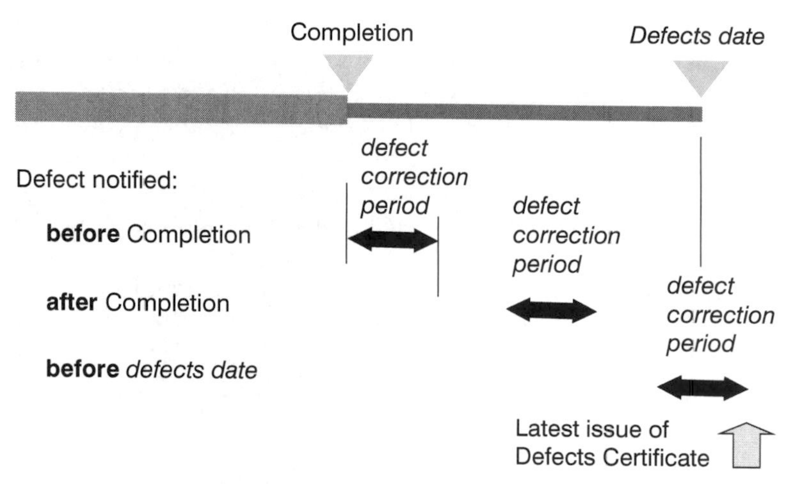

Figure 4.2 Defects and the Defects Certificate

Note that the Contractor's liability for correction of latent defects and other costs associated with them will be dependent on the applicable law, and liability may remain despite the issue of the Defects Certificate.

Question 4.6 Our project, which we are about to carry out using the NEC3 Engineering and Construction Contract Option B (priced contract with bill of quantities), will be on completion a very high security building with restricted access. Once the Employer has taken over the works, we cannot allow the Contractor to return to correct any defects. How do we then deal with any defects which may arise?

Normally in a construction contract, the Contractor is obliged to correct defects arising in the works, during the works and following Completion, in the case of NEC3 contracts until the defects date. This is an obligation, but also it gives the Contractor the opportunity to correct any defects.

However, there may be a situation where the Employer cannot allow the Contactor back into the works due to high security e.g. nuclear projects.

Under Clause 45.2 of the Engineering and Construction Contract, if the Contractor is not given access in order to correct a notified Defect before the *defects date*, the Project Manager assesses the cost to the Contractor of correcting the Defect and the Contractor pays this amount. The Works Information is then treated as having been changed to accept the Defect.

This is an unusual clause in that most contracts as stated provide for the Contractor to correct his own defects; if he does not take the opportunity he is liable for the cost of someone else correcting it.

In this case, as the Contractor is denied the right to re-enter the building he is subject to the Project Manager's subjective opinion as to how much cost the Contractor would have incurred in correcting the Defect(s) if he had been allowed access.

Whilst this may seem a simple and effective solution, in practical terms it is very difficult for the Project Manager to assess the cost to the Contractor of correcting the Defect.

Question 4.7 The Contractor has installed the wrong suspended ceiling tiles to the training rooms within our new building. However, we urgently need to take over the works and cannot wait for him to correct the defect. How can we manage this issue?

A Defect is defined as "a part of the works that is not in accordance with the Works Information" or "a part of the works designed by the Contractor which is not in accordance with the applicable law or the Contractor's design which the Project Manager has accepted" (Clause 11.2(5)).

Example

The Contractor has carried out a large area of wall tiling in a proposed new railway station. Whilst visually the quality of the work appears to be very good, the tiles have not been laid to the required tolerance within the Works Information and therefore the work is defective.

The railway station is due to be completed in two weeks, so if the Contractor has to take down the wall tiles, re-order another batch of tiles and then lay the new tiles to the required tolerance, that could take a considerable time and possibly prevent completion of the works or that specific part of the works.

In this case, provided both the Contractor and the Project Manager are prepared to consider changing the Works Information, then the Contractor provides a quotation to the Project Manager stating a financial saving (reduced Prices), based on him not having to correct the Defect or an earlier Completion Date or both, and if the Project Manager accepts the quotation then the Works Information is changed and the Prices and/or Completion Date are changed in accordance with the quotation.

This is obviously not an option where the design is not in accordance with the applicable law. If the Contractor and Project Manager consider this change, the Contractor submits a quotation for reduced Prices or an earlier Completion Date, or both.

If the Project Manager accepts the quotation he gives an instruction to change the Works Information, the Prices and the Completion Date.

In the event that a Defect becomes apparent, there are two choices.

1 The Contractor corrects the Defect so that the part of the Works is in accordance with the Works Information or the design is in accordance with the applicable law or the Contractor's design.
2 The Contractor and Project Manager may each propose to the other that the Works Information should be changed so that a Defect does not have to be corrected. Note that "each proposes to the other" requires the acceptance of the Contractor and the Project Manager, the latter probably discussing the matter with the Employer.

Chapter 5

Payment provisions

Payment and non-payment under the various options, use of Disallowed Cost, etc.

Question 5.1 We wish to make use of a Project Bank Account. Can we do this with an NEC3 contract?

In 2008, the Office of Government Commerce (OGC) published a guide to fair payment practices, following which the NEC Panel prepared a document in June 2008 to allow users to implement these fair payment practices in NEC contracts.

This document originally entitled "Z3: Project Bank Account", now Option Y(UK)1, authorises a Project Bank Account which, if we use the Engineering and Construction Contract as an example, receives payments from the Employer which is in turn used to make payments to the Contractor and to Named Suppliers.

There is also a Trust Deed between the Employer, the Contractor and Named Suppliers containing the necessary provisions for administering the Project Bank Account. This is executed before the first assessment date.

Additionally, the Contractor includes in his subcontracts for Named Suppliers to be party to the Project Bank Account through a Trust Deed. The Contractor notifies the Named Suppliers of the details of the Project Bank Account and the arrangements for payment of amounts due under their contracts.

The Named Suppliers will be named within the Contractor's tender, and additional Named Suppliers may be included subject to the Project Manager's acceptance by means of a Joining Deed, which is executed by the Employer, the Contractor and the new Named Supplier. The new Named Supplier then becomes a party to the Trust Deed.

As the Project Bank Account is maintained by the Contractor, he pays any bank charges and also is entitled to any interest earned on the account. The Contractor is also required at tender stage to put forward his proposals for a suitable bank or other entity which can offer the arrangements required under the contract.

The process every month is that, at each assessment date, the Contractor submits an application for payment to the Project Manager including details of amounts due to Named Suppliers in accordance with their contracts.

No later than one week before the final date for payment, the Employer makes Payment to the Project Bank Account of the amount which is due to be paid to the Contractor. If the Project Bank Account has insufficient funds to make all the payments, particularly to the Named Suppliers, the Contractor is required to add funds to the account to make up the shortfall.

The Contractor then prepares the Authorisation, setting out the sums due to Named Suppliers. After signing the Authorisation, the Contractor submits it to the Project Manager for signature by the Employer and submission to the Project Bank.

The Contractor and Named Suppliers then receive payment from the Project Bank Account of the sums set out in the Authorisation after the Project Bank Account receives payment.

In the event of termination, no further payments are made into the Project Bank Account.

Question 5.2 We note that the Engineering and Construction Contract does not include provision for retention on payments to the Contractor. Why is this, and how can we include for retention?

It is true that retention is not included within the core clauses of the Engineering and Construction Contract, but it is included as a Secondary Option – i.e. Option X16.

This is because not all clients will want to deduct retention from payments; therefore it is an option, rather than a core clause requirement.

If Secondary Option X16 is selected, the Employer may retain a proportion of the Price for Work Done to Date once it has reached any retention free amount, the retention percentage and any retention free amount being stated in Contract Data Part 1.

Note that, under Clause 11.2(23) (Options C, D, E and F), Defined Cost does not include amounts deducted for retention. So, if the Contractor deducts retention from a Subcontractor, the figure before the deduction is used in calculating Defined Cost, ensuring that there is no "double deduction" of retention.

Following Completion of the whole of the works, or the date the Employer takes over the whole of the works, whichever happens earlier, the retention percentage is halved, and then the final release is upon the issue of the Defects Certificate.

It is important to note that retention is held against undiscovered defects and not incomplete work.

Option X16 is included within the Engineering and Construction Contract.

Retention is provided within the core clauses in the Engineering and Construction Short Contract.

Question 5.3 We need to pay the Contractor on an NEC3 Engineering and Construction Contract partly in US dollars and partly in the local currency of the country in which the project is located. How can we do this?

It is a fairly common requirement in international contracts for part of the payments to the Contractor or the Consultant to be in, for example, the US dollar, which may be linked to oil prices, and part in the local currency of the country in which the project is located.

The currency of the contract is stated in Contract Data Part 1. For example, the project may be in Dubai, and the currency of the contract would most probably be the Dubai Dirham, in which case all payments would be made in Dirhams.

However, Secondary Option X3 provides for specified items or activities to be paid in an alternative currency, the items or activities, the currency of the payment and the total maximum payment in this currency to be listed in the Contract Data, beyond which payments are made in the currency of the contract.

The exchange rates, their source and date of publication are also referred to in the Contract Data.

Option X3 is included within:

- the Engineering and Construction Contract
- the Professional Services Contract
- the Term Service Contract.

Question 5.4 We wish to pay the Contractor on an NEC3 Engineering and Construction Contract a mobilisation payment equating to 10 per cent of the value of the contract. How can we do this?

Mobilisation payments are appropriate when the Contractor will incur significant "up front" costs before he starts receiving payments – for example, in pre-ordering specialist Materials, Plant or Equipment. The payment is made within four weeks of the Contract Date or if an Advanced Payment Bond is required at the later of the Contract Date and the date the Employer receives the Bond. If the advanced payment is delayed it is a compensation event.

If an Advanced Payment Bond is required it is issued by a bank or insurer which the Project Manager has accepted, the Bond being in the amount that the Contractor has not repaid. Advance Payment Bonds are a helpful security when an advance payment is made to a Contractor for works to be performed. The Project Manager must accept the provider of that bond.

The amount of the payments is stated in the Contract Data Part 1 together with the repayment amounts and any requirements for a bond or security.

Where the advance payment reduces with time as, for example, stage payments are made against goods and/or services delivered under the contract, then the value of the Bond should reduce to reflect the outstanding amount of the advance payment.

Question 5.5 What is the Fee in an NEC3 Engineering and Construction Contract intended to cover?

Fee is defined under Clause 11.2(8) as "the sum of the amounts calculated by applying the subcontracted fee percentage to the Defined Cost of subcontracted work and the direct fee percentage to the Defined Cost of other work".

Fee is therefore an amount calculated by applying the relevant Fee percentage to the relevant Defined Cost.

Whilst many NEC3 practitioners believe that the Fee within the Engineering and Construction Contract simply covers "overheads and profit", the definition is a little wider.

The Fee covers any financial amount, whether it be cost or profit that is not listed in the Schedules of Cost Components.

There is no schedule of items covered by the Fee, but the following list, whilst not exhaustive, gives some examples of cost components not included in the Schedules of Cost Components.

1 Profit.
2 The cost of offices outside the Working Areas, e.g. the Contractor's head office.
3 Insurance premiums.
4 Performance bond costs.
5 Corporation tax.
6 Advertising and recruitment costs.
7 Sureties and guarantees required for the contract.
8 Some indirect payments to staff.

From this one can see that there are some elements covered by the Fee percentage which could be mistakenly assumed as being covered by the Schedules of Cost Components.

Previous editions of the Engineering and Construction Contract had only one fee percentage applicable to total cost, but the NEC3 Engineering and Construction Contract allows the Contractor to tender two fee percentages:

1 the "subcontracted fee percentage" applied to the Defined Cost of subcontracted work
2 the "direct fee percentage" applied to the Defined Cost of other work.

It is essential that these two fee percentages are correctly allocated to the appropriate costs when assessing payments and compensation events.

In reality, many Contractors tend to bracket the two fee percentages together as a single fee percentage, which they are entitled to do, and this in effect makes life easier for the Contractor and the Project Manager when assessing payments and compensation events.

Note that under Options A and B, Defined Cost is the cost of components in the Shorter Schedule of Cost Components whether work is subcontracted or not, whereas under Options C, D and E, where the Schedule of Cost Components is normally used, reference to the Contractor means the Contractor and not his Subcontractors. Subcontractors' costs are dealt with separately.

The percentage for Fee is used for:

1 payments under Options C to F – i.e. Defined Cost forecast to have been paid by the next assessment date plus Fee, and
2 in the assessment of compensation events, i.e. the changes to the Prices are assessed as the effect of the compensation event upon:
   - the actual Defined Cost of work already done
   - the forecast Defined Cost of the work yet to be done, and
   - the resulting Fee.

That said, if the Project Manager *and* Contractor agree, rates and lump sums may be used instead of Defined Cost.

Question 5.6 What is the Overheads and Profit percentage in an NEC3 Engineering and Construction Short Contract intended to cover?

Unlike the NEC3 Engineering and Construction Contract, the Engineering and Construction Short Contract does not include the term "Fee" nor does it have Schedules of Cost Components.

It has two percentages for overheads and profit:

1 the percentage for overheads and profit added to the Defined Cost for people
2 the percentage for overheads and profit added to other Defined Cost.

By reference to Clause 11.2(5), Defined Cost is the amount paid by the Contractor in Providing the Works (excluding any tax which the Contractor can recover) for:

- people employed by the Contractor
- Plant and Materials
- work subcontracted by the Contractor, and
- Equipment.

As with the Engineering and Construction Contract, it is not specific as to what the percentage for overheads and profit covers, but again it would typically include the following.

1 Profit.
2 The cost of offices outside the Working Areas, e.g. the Contractor's head office.

3 Insurance premiums.
4 Performance bond costs.
5 Corporation tax.
6 Advertising and recruitment costs.
7 Sureties and guarantees required for the contract.
8 Some indirect payments to staff.

From this one can see that there are some elements covered by the Fee percentage which could be mistakenly assumed as being covered as Defined Cost.

The percentages for overheads and profit are used when assessing compensation events.

Under the Engineering and Construction Short Contract, there are two methods of pricing compensation events.

1 For compensation events which only affect the quantities of work shown in the Price List: the change to the Prices is assessed by multiplying the changes quantities by the appropriate rates in the Price List.
2 For other compensation events: the change to the Prices is assessed by forecasting the effect of the compensation event upon the Defined Cost or if the compensation event has already occurred, due to the event has occurred, plus the percentage for overheads and profit.

Effects are assessed separately for the following.

- People employed by the Contractor: this relates to the cost of People who are directly employed by the Contractor, excluding Subcontractors, and must include management as well as operatives. The cost of design should also be considered, though if that is subcontracted it would be covered below.
- Plant and Materials: these are "items to be included in the works". The cost of delivery, providing and removing packaging, unloading, etc. should be included. The cost of disposal of Plant and Materials should be credited.
- Work subcontracted by the Contractor: subcontracted design should also be considered.
- Equipment: equipment is "items provided by the Contractor, used by him to Provide the Works and not included in the works", so this is what other contracts refer to as Plant. This would include

excavators, cranes, scaffold and temporary accommodation. The cost of transporting Equipment to and from the Site and the erection and dismantling of the Equipment should be included.

- Note that the amount for Equipment includes amounts paid for hired Equipment and an amount for the use of Equipment owned by the Contractor which is the amount the Contractor would have paid if the Equipment had been hired.

Defined Cost is defined in Clause 63.3 as open market or competitively tendered prices with deductions for all documents, rebates and taxes deducted.

In addition the following should be deducted from Defined Cost:

- the cost of events for which the contract requires the Contractor to insure, and
- other costs paid to the Contractor by insurers.

Question 5.7 We are confused by the two terms "Working Areas Overheads" and "People Overheads" in the NEC3 Engineering and Construction Contract. What do these terms mean and how are they calculated?

The term "Working Areas overheads" is included within the Schedules of Cost Components and the term "People Overheads" is included within the Shorter Schedule of Cost Components.

Schedule of Cost Components

Clause 4: Charges

This covers various miscellaneous costs incurred by the Contractor such as temporary water, gas and electricity, payments to public authorities, and also payments for various other charges such as cancellation charges, buying or leasing of land, inspection certificates and facilities for visits to the Working Areas.

The cost of any consumables and equipment provided by the Contractor for the Project Manager's or the Supervisor's offices is also included as direct cost. (Note the cost of the Contractor's own equipment is covered in the Equipment Section, and the Contractor's consumables are included within the Working Areas overheads percentage.)

Item 44 covers various Site consumables as listed from (a) to (j).

Table 5.1 **The Schedule of Cost Components**

Item 44	*Included*	*Not Included*	*Where is it included*
(a) Catering	Catering equipment Food Electricity Bills	Canteen accommodation Catering staff	Equipment People
(b) Medical facilities	First aid box Stretchers	First aid personnel First aid accommodation	People Accommodation
(c) Recreation	Television Radio Gym Facilities		
(d) Sanitation	Toilet rolls Hand towels Soap Barrier creams	Toilet Accommodation	Equipment
(e) Security	Security systems Alarms	Security personnel Guard dogs	People Equipment
(f) Copying	Photocopier Photocopier toner and paper	Other stationery	Equipment
(g) Telephone, telex, fax, radio and CCTV	Telephones Fax Machine		
(h) Surveying and setting out	Surveying equipment	Site Engineer Surveyors	People
(i) Computing	Hardware Software Printers Print paper	IT Personnel	People
(j) Hand tools not powered by compressed air	Electric tools		

It is important to recognise that Item 44 only includes for provision and use of equipment, supplies and services, but excludes accommodation, so if we consider the list (see Table 5.1), it is important to recognise what is and is not included.

The cost is calculated by multiplying the Working Areas overheads percentage inserted by the Contractor in Contract Data Part 2 by the People cost items 11, 12, 13 and 14.

Shorter Schedule of Cost Components

Clause 4: Charges

This covers various miscellaneous costs incurred by the Contractor such as temporary water, gas and electricity payments to public authorities, and also payments for various other charges, which may or may not be relevant dependent on the project.

These costs are not calculated on a direct cost basis but by reference to the percentage for People overheads which is applied to People costs.

Question 5.8 The Contractor on an NEC3 Engineering and Construction Contract has failed to submit a programme to the Project Manager. What can we do to remedy this?

First, in respect of payments due to the Contractor, if no programme is identified in Contract Data Part 2 (i.e. it was submitted with the Contractor's tender), one-quarter of the Price for Work Done to Date is retained (Clause 50.3) until the Contractor has submitted the first programme showing the information the contract requires.

Note that a Contractor who submits a first programme which shows the required information, but the Project Manager does not accept it, would not be liable for withholding of payment under this clause. The clause relates to the Contractor's failure to submit a programme showing the information the contract requires, not the acceptance of it.

Second, in respect of compensation events, the Project Manager may assess a compensation event himself if:

- when the Contractor submits quotations for a compensation event, he has not submitted a programme which the contract requires him to submit, or
- when the Contractor submits quotations for a compensation event, the Project Manager has not accepted the Contractor's latest programme for one of the reasons stated in the Contract.

Question 5.9 On an NEC3 Engineering and Construction Contract Option A (priced contract with activity schedule), if the Contractor has completed an activity, should he be paid for that activity if it contains Defects?

The Price for Work Done to Date (PWDD) under Option A is the total of the Prices for completed activities. A completed activity is one without Defects which would "either delay or be covered by immediately following work".

So, for example, if the activity was to install a 30-metre length of copper pipework on and including supports, attached to a wall, and the pipework and the supports were all installed but there was a Defect in one of the supports, then provided it does not delay or is covered by immediately following work, it is a completed activity, but with a Defect that needs to be corrected.

The Employer paying for the activity because the Project Manager assesses that it is completed is not the Employer accepting the work or the Defects within it, it is just a means to create cash flow to the Contractor, the Contractor is still obliged to correct the Defect and is also responsible if, for example, someone damages the work once it is paid for.

Conversely, if the pipework was a drain pipe in a trench and there is a Defect which will be covered up the following day when the Contractor backfills the trench, then it is not paid at the time of assessment, the Contractor must correct the Defect first.

The Contractor is not entitled to be paid for activities which are not complete or are only partly complete, so the decision always has to be made by the Project Manager at the assessment date whether the activity is either complete or not complete. Assessing the Contractor's right to payment under the contract is therefore different to the traditional approach of valuing the works carried out to date.

Question 5.10 We are Contractors tendering for a series of NEC3 Engineering and Construction Contracts using Option A (priced contract with activity schedule). How should we price our Preliminaries costs within the activity schedule so that we are paid correctly?

It is not uncommon for Contractors who are new to NEC3 contracts, when they submit tenders for an NEC3 Engineering and Construction Contract Option A (priced contract with activity schedule), to include their Preliminaries costs as a single sum of money within a single activity believing they will actually be paid these costs in a traditional way – i.e. in the first month a lump sum payment for setting set up his Site establishment, in continuing months payments for maintaining Site as a time-related sum, then in the final month a lump sum payment for clearing the Site. However, this is an incorrect assumption.

The Price for Work Done to Date (PWDD) under Option A is the total of the Prices for completed activities. Therefore, the Contractor should consider breaking down his Preliminaries as follows.

1 **Fixed preliminaries:** These would comprise Preliminaries which may be single items, not directly influenced by progress on Site.

   Typical examples are delivery of temporary accommodation and other equipment to Site, connection of telephones and other temporary services.

   These should be identified within the Activity Schedule as single activities, e.g. "delivery of Site accommodation", or a group of activities identified as "set up Site".

   When the activities are completed, they are included within the next assessment following completion.
2 **Time related preliminaries:** These comprise Preliminaries which are based on time on Site rather than having any direct relationship to the quantity of work carried out.

   Typical examples are site management salaries, site accommodation hire charges, and scaffolding.

   These should be identified within the Activity Schedule as time-based activities, e.g. "one month hire of temporary accommodation". Again, when the activities are completed, they are included within the next assessment.

On a similar theme Contractors who are tendering for NEC3 Engineering and Construction Contracts Option A (priced contract with activity schedule), where they have design responsibility, often include their Design costs as a single sum.

The Contractor should not just include an activity titled "Design" as there are many aspects of design, and one could argue that design is not completed until the project is completed, so he could deny himself payment for a considerable period.

Ideally, the design related activities should be linked to various deliverables rather than generic activity descriptions such as "prepare drawings".

The various elements of design could be related to, for example, the various stages within the RIBA Plan of Work, or in its simplest form there could be separate activities for the design of various elements of the project, and for the granting of planning approval, etc.

Question 5.11 On an NEC3 Engineering and Construction Contract Option B (priced contract with bill of quantities), the Contractor contends that if the quantities change, he is entitled to revise his rates. Is that true?

No, that is incorrect. Option B is a remeasurement contract, if the quantities change and it is not because of a compensation event, the original rates apply.

However, within Option B, one must also consider the following clauses.

Clause 60.4: "A difference between the final total of work done and the quantity for an item in the Bill of Quantities is a compensation event if:

- the difference does not result from a change to the Works Information
- the difference causes the Defined Cost per unit of quantity to change
- the rate in the Bill of Quantities for the item multiplied by the final total quantity of work done is more than 0.5 per cent of the Total of the Prices at the Contract Date."

A difference between the final total quantity of work done and the quantity for an item on the Bill of Quantities is not a compensation event in itself, unless it satisfies all three bullet points within the clause.

Let's examine each of the bullet points in turn.

- the difference does not result from a change to the Works Information

If it had resulted from a change to the Works Information, then it would have been dealt with as a compensation event under Clause 60.1(1).

- the difference causes the Defined Cost per unit of quantity to change

Example

Original quantity of work Road Gullies 150 @ £125 = £18,750

Due to an error in measuring the road gullies when preparing the Bill of Quantities, the final quantity is only 120.

Final quantity of work Road Gullies 120 @ £125 = £15,000

The Total of the Prices at the Contract Date is £2,000,000.

0.5% × £2,000,000 = £10,000. Therefore, assuming that all three conditions are satisfied, then it is a compensation event.

The purchase price of materials may be affected by a change in the quantity, for example a price reduction or increase may be applicable when the number of units exceeds or falls below a certain amount. Or the cost of mobilisation or delivery could have changed as the same vehicle delivers fewer units.

- the rate in the Bill of Quantities for the item multiplied by the final total quantity of work done is more than 0.5 per cent of the Total of the Prices at the Contract Date

This final bullet point compares the final extended price for that item in the Bills of Quantities with the Total of the Prices at the Contract Date (many contracts refer to this as the "Tender Sum" or the "Contract Sum"). If the extended price is more than 0.5 per cent of the Total of the Prices at the Contract Date, and both of the previous bullet points are satisfied, then it is a compensation event. This final bullet excludes items of a "minor value" in the Bill of Quantities.

Question 5.12 What are "accounts and records" as required under an NEC3 Engineering and Construction Contract Option C (target contract with activity schedule)? The Contractor has stated that there are certain accounts and records of costs that he cannot make available to the Project Manager as they are confidential, and also in some cases would breach data protection law. How can we pay the Contractor these costs?

The Contractor is required to keep the following accounts and records (Clause 52.2) to calculate Defined Cost:

- accounts of payments of Defined Cost
- proof that the payments have been made
- communications about and assessments of compensation events for Subcontractors, and
- other records as stated in the Works Information.

What would these records consist of? Essentially, all that the Contractor requires to prove every cost for which he is seeking reimbursement under the contract i.e. invoices, orders, timesheets, receipts, etc.

The level of checking of the Contractor's accounts and records is at the Project Manager's discretion, some wish to examine each cost and its relevant back-up document, some wish only to select certain costs at random, others wishing to only carry out a cursory check of costs.

Some Project Managers say that if the parties really are acting in a spirit of mutual trust and co-operation as the contract requires, then there should be no need for a detailed examination of the Contractor's accounts, though this probably needs a quantum leap of faith for many Project Managers and Employers!

It is important to establish the format on which accounts and records are to be presented by the Contractor to the Project Manager, particularly in respect of assessing payments, and this can be detailed in the Works Information, the following alternative methods normally being employed:

1 The Contractor submits a copy of all of his accounts and records to the Project Manager on a regular basis.
2 The Contractor gives the Project Manager access to his accounts and records and the Project Manager takes copies of all the records he requires. Whilst not expressly stated within the

contract, the Project Manager should not have to travel too far to inspect them. It has been known for accounts and records to be "available for inspection" in a different country to the Site, the Contractor stating that they were available for inspection at any time during normal working hours! Again, the location and availability of accounts and records could be detailed within the Works Information.

3 The Contractor gives the Project Manager direct access to his computerised accounts and records through a password or PIN, the Project Manager can then take copies of the records he requires.

Let's now consider the payment process within Option C.

The Price for Work Done to Date is the Defined Cost which the Project Manager forecasts will have been paid by the Contractor before the next assessment date plus the Fee, so the assessment is a combination of Defined Cost paid and yet to be paid by the Contractor.

For example, if the assessment date is 1 June, then costs which are forecast to be paid by 1 July are also included.

This is a change to previous editions of the contract which provided for the Contractor only to be paid cost that he had paid at the assessment date, the change being introduced to assist the Contractor's cash flow.

Whilst the forecast may appear to be no more than a subjective guess, the Contractor has an up to date programme and also has probably received many of the invoices which he will be paying before the next assessment date.

In practice, however, many Employers are resistant to forecasting costs a month ahead and therefore tend to amend the contract through a Z clause to retain the previous wording.

In assessing the amount due if the Contractor has paid costs in a currency different to the currency of the contract, the Contractor is paid Defined Cost in the same currency as the payments made by him. All payments are converted to the currency of the contract applying the exchange rates, in order to calculate the Fee.

In response to the question that the Contractor has stated that there are certain accounts and records of costs that he cannot submit to the Project Manager as they are confidential, and also in some cases would breach data protection law, the simple answer is that he cannot be paid these costs as the Project Manager is unable to forecast that they will have been paid by the Contractor before the next assessment date as he has not been able to verify what these costs are.

The Contractor may feel that salaries in particular are confidential or sensitive, but he must make the accounts and records available to the Employer in order to be paid them, but in turn the Project Manager, as with all the accounts and records made available to him, must respect that confidentiality.

Question 5.13 What are Disallowed Costs and how are they deducted from payments due? Can Disallowed Cost be applied retrospectively to a payment made in a previous month? Is there a maximum timescale for deducting Disallowed Cost?

Disallowed Costs are costs which would normally be payable to the Contractor, but are disallowed under certain clauses within the contract.

There are different clauses covering Disallowed Cost dependent on the selected Main Option.

Under Options C, D and E, the relevant Disallowed Cost clause for all three Options is Clause 11.2(25).

Disallowed Cost is cost which the Project Manager decides:

- is not justified by the Contractor's accounts and records. The Contractor is obliged to keep accounts, proof of payments, communications regarding payments, and any other records as stated in the Works Information (Clauses 52.3 and 52.3). If he does not have the accounts and records to prove a cost then the cost must be disallowed. It is not sufficient to merely prove that the goods or materials are on the Site for the Project Manager to see, or held off site at a designated place for inspection, as with many other contracts.
- should not have been paid to a Subcontractor or supplier in accordance with his contract. The Contractor is obliged to submit the proposed conditions of contract for each Subcontractor to the Project Manager for acceptance, unless an NEC contract is proposed, or the Project Manager has agreed that no submission is required. In addition under Options C, D, E and F, the Contractor is required to submit the proposed Contract Data for each subcontract if an NEC contract is proposed and the Project Manager instructs the Contractor to make the submission. If the Contractor pays a Subcontractor or supplier an amount that is not in accordance with his contract with them, then this is disallowed.
- was incurred only because the Contractor did not:

- follow an acceptance or procurement procedure stated in the Works Information. The Works Information may, for example, require the Contractor to comply with a specific procedure in respect of, for example, submission of design proposals or procurement of Subcontractors. If the Contractor does not comply with the procedure then associated costs are disallowed.
- give an early warning which the contract required him to give. If the Contractor incurs a cost that could have been avoided if the Contractor had given early warning, then it is disallowed.

- correcting Defects after Completion. This is a Disallowed Cost. However, correction of Defects before Completion is not a Disallowed Cost. This is often a contentious issue with Employers and Project Managers objecting to paying the Contractor for correcting his own Defects! However, one must consider this clause logically. The Contractor has quite possibly priced the risk of having to correct Defects which are his liability within his original tender, on an Option C contract, for example, this price will be included within the target. Option C is a cost reimbursable contract until Completion, and the only way the Contractor can be paid for this risk, should it materialise, is through the Defined Cost process as Option C is a cost reimbursable contract until Completion. It is not a compensation event, so the target and/or the Completion Date are not changed. The Contractor is correcting the Defect in his own time and potentially reducing the share he could make at Completion.
- correcting Defects caused by the Contractor not complying with a constraint on how he is to Provide the Works stated in the Works Information. A constraint may be stated in the Works Information, such as a prescribed method of working. For example, the Works Information may prescribe that a hardcore sub-base filling must be rolled six times with a vibrating roller of a certain weight, but the Contractor does not do so, and later a defect arises due to the Contractor's failure to comply with that stated constraint. If the Contractor does not comply with this constraint and a Defect occurs, then the Contractor's cost of correcting the Defect is disallowed.
- Plant and Materials not used to Provide the Works (after allowing for reasonable wastage) unless resulting from a change to the Works Information. Excess wastage of Plant or Materials beyond what is considered reasonable is a Disallowed Cost. The question of what is "reasonable" can often be debatable! As with all

Disallowed Cost, it is the Project Manager's responsibility to make the decision and to disallow the cost.

- resources not used to Provide the Works (after allowing for reasonable availability and utilisation) or not taken away from the Working Areas when the Project Manager requested. This provision will include People and Equipment. If the Contractor does not remove Equipment when it is no longer required, then this is Disallowed Cost. If the Contractor is using more resources than he has planned and priced for, or his resources are inefficient, that is not a Disallowed Cost.
- preparation for and conduct of an adjudication or proceedings of the tribunal. If an adjudication, arbitration or legal proceedings occur, then each party bears his own costs.

This final bullet was not introduced until NEC3 was published, so prior to NEC3 a Contractor could, in theory, refer a dispute to adjudication and, whether or not he was successful, any costs arising could be included as cost and would not be disallowed.

Under Option F, the relevant Disallowed Cost clause is Clause F11.2.(26).

Disallowed Cost is cost which the Project Manager decides:

- is not justified by the Contractor's accounts and records
- should not have been paid to a Subcontractor or supplier in accordance with his contract
- was incurred only because the Contractor did not:
  - follow an acceptance or procurement procedure stated in the Works Information
  - give an early warning which the contract required him to give
- is a payment to a Subcontractor for:
  - work which the Contractor states that the Contractor will do himself, or
  - the Contractor's management.

Clearly, with all the Disallowed Cost provisions, as the Project Manager is to decide that a cost is to be disallowed, then he has to be diligent enough to identify the cost, to quantify it, and to make the appropriate deductions, which are not the easiest of tasks!

In answer to the question "Can Disallowed Cost be applied retrospectively to a payment made in a previous month?", the simple answer is yes, Disallowed Cost can be deducted from any payment which would otherwise be due to the Contractor and does not necessarily have to be within the current payment to which the Disallowed Cost relates.

Question 5.14 How is the Contractor's Share under the NEC3 Engineering and Construction Contract Option C (target contract with activity schedule) set and calculated?

The Contractor tenders a price and includes an activity schedule in the same way as he would under an Option A contract. This price, when accepted, is then referred to as the "target".

The Contractor also tenders his percentage(s) for Fee. The original target is referred to as the "Total of the Prices at the Contract Date".

- The target price includes the Contractor's estimate of Defined Cost plus other costs, overheads and profit to be covered by his Fee.
- The Contractor tenders his Fee in terms of percentages to be applied to Defined Cost.
- During the course of the contract, the Contractor is paid Defined Cost plus the Fee.
- The target is adjusted for compensation events and also for inflation (if Option X1 is used).
- On Completion, the Project Manager assesses the Contractor's share in accordance with Clause 53.1 which, it has to be said, is at best a confusing clause, though the Guidance Notes clarify the clause! The Contractor then pays or is paid his share of the difference between the final total of the Prices and the final Payment for Work Done to Date according to a formula stated in the Contract Data.
- This motivates the Contractor to decrease costs. Many refer to this sharing of risk and opportunity as "pain and gain". It often comes as a surprise to NEC users that the terms "pain and gain" do not appear anywhere within the NEC contracts, nor do the terms "target price" or "target cost"!

Question 5.15 How are unfixed materials on or off Site dealt with under the NEC3 contracts?

A common area of confusion is the payment to the Contractor for unfixed materials within or outside the Working Areas, i.e. materials on or off Site. Unlike other forms of contract, the NEC3 has no express provisions for payment for such materials.

Within the Engineering and Construction Contract, for example, in order for the Contractor to be paid for unfixed materials, whether they be on or off Site, he should consider the payment rules for each of the Main Options.

Option A: As the Price for Work Done to Date (PWDD) is based on completed activities, the Contractor should create an activity in the activity schedule for materials. For example, for a structural steel frame, the Contractor could include four activities:

1 Delivery of steel to Site
2 Erection of steel to Gridline 1–10
3 Erection of steel to Gridline 10–20
4 Paint steel.

As each activity is completed, the Contractor is then paid within the assessment following completion of each activity.

Option B: As the PWDD is based on Bills of Quantities, appropriate items may be included as method related charges.

Option C, D, E and F: As the PWDD is the Defined Cost which the Project Manager forecasts will have been paid by the Contractor before the next assessment date plus the Fee, the Contractor must provide accounts and records to show that he has paid for the materials or will have paid for them by the next assessment date.

Note that, in respect of unfixed materials on Site, whatever title the Contractor has to Plant and Materials passes to the Employer if it has been brought within the Working Areas and passes back to the Contractor if it is removed from the Working Areas with the Project Manager's permission.

In respect of unfixed materials off Site, whatever title the Contractor has to Plant and Materials passes to the Employer if the Supervisor has marked it as for this contract.

Also, in respect of marking, the contract must have identified them for payment, and the Contractor must have prepared them for marking as required by the Works Information. This could include setting them aside from other stock, protection, insurance and any vesting requirements.

Question 5.16 How is interest calculated on late payments on an NEC3 Engineering and Construction Contract?

Interest is paid on late payments, the interest rate being stated in Contract Data Part 1 and assessed on a daily basis from the date the payment should have been made until the date when the late payment is made, calculated using the interest rate in Contract Data Part 1 compounded annually.

The interest due can be calculated on the basis of the following formula:

Payment due × interest rate × the number of days late / 365 days

So if one assumes the following:

Payment due = £120,000

Payment seven days late

Interest rate in Contract Data Part 1 = 3%

the calculation is:

£120,000 × 3% × 7 days / 365 days Interest due = £69.04

Similarly, interest paid on a correcting amount due applies to later corrections to certified amounts by the Project Manager and on interest due to a compensation event or as determined by the Adjudicator or the tribunal. The date from which interest should run is the date on which the additional payment would have been certified if there had been no dispute or mistake.

Question 5.17 The Contractor on an NEC3 Engineering and Construction Contract has proposed that if the design of some structural elements of the project are changed from in situ concrete to precast concrete he could offer cost and time savings. How would this be dealt with under the contract?

The answer to this question lies within Clause 63.11 (Bullet Point 1):

"If the effect of a compensation event is to reduce the total Defined Cost and the event is:

- a change to the Works Information, other than a change to the Works Information provided by the Employer which the Contractor proposed and the Project Manager has accepted."

If the Contractor proposes a change to the Works Information as part of a value engineering exercise, and the Project Manager accepts, then the Prices are not reduced i.e. the target is not reduced.

This is to encourage the Contractor to initiate value engineering solutions which generate cost savings which are then, if the Project Manager accepts, shared between the Employer and the Contractor through the final Contractor's share.

Question 5.18 The Contractor on an NEC3 Engineering and Construction Contract has failed to submit an application for payment by the assessment date. The Project Manager has stated that no payment is therefore due to the Contractor this month. Can we assume that this is correct?

Let us start by considering the payment assessment process.

The first assessment date is decided by the Project Manager to suit the parties, and will normally be based on the time that the Contractor has been on Site, the Employer's procedures and timing for processing and issuing payments, and the Contractor's payment requirements and internal accounting system.

Clearly, in this respect, some discussion needs to take place between the Project Manager, the Employer and the Contractor in order that a convenient assessment date can be set.

The first assessment must be made within the "assessment interval" after the starting date, this is normally inserted in the Contract Data as "four weeks" or "one calendar month".

Later assessment dates occur:

- at the end of each assessment interval until four weeks after the Supervisor issues the Defects Certificate
- at Completion of the whole of the Works.

There is no provision within the contract for a minimum certificate amount. The Project Manager assesses the amount due at each assessment date, calculating the Price for Work Done to Date (PWDD) using the rules of the specific Main Option. The amount due to the Contractor is:

- the PWDD
- **plus** other amounts to be paid to the Contractor (e.g. Contractor's share, bonus for early completion, value added tax)
- **less** amounts to be paid by or retained from the Contractor (e.g. retention, delay damages).

In making his assessment, the Project Manager considers "any application for payment the Contractor has submitted". Note that in using the phrase "any application", the Contractor does not have to submit an application.

If the Contractor submits an application, the Project Manager considers that, if the Contractor does not the Project Manager must still make the assessment based on the requirements of the specific Main Option chosen.

It is very unlikely that the Contractor will not submit an application for payment, but this clause does cause some concern among Employers and Project Managers as if the Contractor does not submit an application for payment and the Project Manager makes a mistake in making the assessment and has to correct it, interest on the correcting amount is paid.

Note: If the correcting amount is an addition, the Employer pays the interest; if a deduction, the Contractor pays the interest.

It is therefore not uncommon for Employers to include within the contract the following Z clause which places the onus on the Contractor to make an application for payment, not on the Project Manager to make the assessment whether or not the Contractor has done so:

"The Contractor submits an application for payment one week before each assessment date. In assessing the amount due, the Project Manager considers the application for payment the Contractor submits. The Project Manager gives the Contractor details of how the amount due has been assessed."

Further assessment dates occur at the end of each assessment interval until four weeks after the Supervisor issues the Defects Certificate, and at Completion of the whole of the works. It is essential that the Contractor either submits a first programme with his tender or within the timescale specified within the contract. Failure to do so will entitle the Project Manager to retain one-quarter of the PWDD in his assessment of the amount due.

Note that the amount is only withheld if the Contractor has not submitted a programme which shows the information which the contract requires, e.g. method statement, provisions for float, etc.

If the Contractor has submitted a programme which contains all the information that the contract requires, but the Project Manager disagrees with, for example, part of the method statement or the programme has not yet been accepted, then the provision does not apply.

Question 5.19 We have a contract under the NEC3 Engineering and Construction Contract where the Contractor has stated that he must be paid for compensation events if he has done the relevant work. Is the Project Manager required to certify payments "on account" for compensation events not yet agreed?

The answer to this question is essentially no, as under Clause 65.4 "the changes to the Prices, the Completion Date and the Key Dates are included in the notification implementing a compensation event". Once implementation occurs, then the compensation event and its effect is incorporated into the contract.

Let's take each Main Option in turn.

Option A: *Priced contract with activity schedule*

The Price for Work Done to Date is defined under Clause 11.2(27).

The Price for Work Done to Date is the total of the Prices for:

- each group of completed activities
- each completed activity which is not in a group.

With Option A, if the compensation event has not yet been implemented and compensation events are in the form of changes to the activity schedule (Clause 63.12), there is no activity as yet, which if complete can be paid.

Option B: *Priced contract with bill of quantities*

11.2(28) The Price for Work Done to Date is the total of:

- the quantity of the work which the Contractor has completed for each item in the Bill of Quantities multiplied by the rate, and
- a proportion of each lump sum which is the proportion of the work covered by the item which the Contractor has completed.

In that case, if the compensation event has not yet been implemented and compensation events are in the form of changes to the Bill of Quantities (Clause 63.13), there is no item in the Bill of Quantities as yet against which payment can be made.

Option C: *Target contract with activity schedule*

11.2(29) The Price for Work Done to Date is the total Defined Cost which the Project Manager forecasts will have been paid by the Contractor before the next assessment date plus the Fee.

With Option C, if the compensation event has not yet been implemented and compensation events are in the form of changes to the activity schedule (Clause 63.12), and the consequent effect on the target, there is no activity as yet against which payment can be made.

However, payment is not based on completed activities, but on Defined Cost, therefore the Contractor can be paid, even though the relevant compensation event has not been implemented.

Option D: *Target contract with bill of quantities*

11.2(29) The Price for Work Done to Date is the total Defined Cost which the Project Manager forecasts will have been paid by the Contractor before the next assessment date plus the Fee.

With Option D, if the compensation event has not yet been implemented and compensation events are in the form of changes to the Bill of Quantities (Clause 63.13), there is no item in the Bill of Quantities as yet against which payment can be made.

However, payment is not based on the Bill of Quantities but on Defined Cost, therefore the Contractor can be paid, even though the relevant compensation event has not been implemented.

Option E: *Cost reimbursable contract*

Option F: *Management contract*

11.2(29) The Price for Work Done to Date is the total Defined Cost which the Project Manager forecasts will have been paid by the Contractor before the next assessment date plus the Fee.

With Option E and F, payment is not based on Defined Cost. Therefore, the Contractor can be paid, even though the relevant compensation event has not been implemented.

Question 5.20 Why is there no provision for Fee, or Overheads and Profit percentages in an NEC3 Professional Services Contract?

Under the NEC3 Professional Services Contract, the Consultant submits staff rates within Contract Data Part 2 identifying the name/designation of the member of staff and the hourly rate.

Then dependent on the Main Option chosen, the Consultant is paid a "Time Charge" which is "the sum of the products of each of the staff rates multiplied by the total staff time appropriate to that rate appropriately spent on work in this contract", so it is essential that these staff rates tendered by the Consultant are inclusive of any costs attributable to those members of staff, and must also include profit as there is no other provision for recovering profit.

Note that the NEC3 Professional Services Contract also provides for expenses to be paid to the Consultant, though this must either be stated by the Employer within Contract Data Part 1 as expenses which the Employer will pay, or by the Consultant within Contract Data Part 2 at the time of tender as expenses which the Consultant wishes to be paid.

The Consultant is required to keep accounts and records of his Time Charge and (Options C, E and G) and, if applicable, expenses (Options A, C, E and G) at any time within working hours.

Under Option A, the Consultant is required to prepare forecasts of the total expenses for the whole of the services on a periodic basis and to submit them to the Employer (Clause 21.3) including an explanation of the changes since the last forecast.

Under Options C, E and G, the Consultant keeps accounts and records of his Time Charge and expenses, and allows the Employer to check them at any time during normal working hours (Clause 52.2). In addition, the Consultant is required to prepare forecasts of the total Time Charge and total expenses for the whole of the services on a periodic basis and to submit them to the Employer (Clause 21.4), including an explanation of the changes since the last forecast.

The level of checking of the Consultant's accounts and records is at the Employer's discretion. Some wish to examine each cost and its relevant back up document, some wish only to select certain costs at random, others wish to only carry out a cursory check of costs. Some Employers say that if the parties really are acting in a spirit of mutual trust and co-operation as the contract requires, then there should be no need for a detailed examination of the Consultant's accounts, though this probably needs a quantum leap of faith for many Employers!

It is important to establish the format on which accounts and records are to be presented by the Consultant to the Employer, particularly in respect of assessing payments, and this can be detailed in the Scope, the following alternative methods normally being employed.

1 The Consultant submits a copy of all of his accounts and records to the Employer on a regular basis.

or

2 The Consultant gives the Employer access to his accounts and records and the Employer takes copies of all the records he requires. Whilst not expressly stated within the contract, the Employer should not have to travel too far to inspect them.

   It has been known for accounts and records to be 'available for inspection' in a different country to the Site, the Consultant stating that they were available for inspection at any time during normal working hours! Again the location and availability of accounts and records could be detailed within the Scope.

or

3 The Consultant gives the Employer direct access to his computerised accounts and records through a PIN, the Employer can then take copies of the records he requires.

Chapter 6

Managing compensation events

Notification, pricing and assessing compensation events, assumptions, etc.

Question 6.1 How do the NEC3 contracts deal with unforeseen ground conditions on the Site?

Engineering and Construction Contract

In considering this question within NEC3 contracts, let us first consider the Engineering and Construction Contract, by reference to Clause 60.1(12):

> "The *Contractor* encounters physical conditions which:
>
> - are within the Site,
> - are not weather conditions, and
> - an experienced contractor would have judged at the Contract Date to have such a small chance of occurring that it would have been unreasonable for him to have allowed for them.
>
> Only the difference between the physical conditions encountered and those for which it would have been reasonable to have allowed is taken into account in assessing a compensation event."

It is important to interpret this final paragraph correctly because if the Contractor did not allow anything in terms of Price and/or time within his tender for dealing with a physical condition he should only be compensated for the difference between what he found and what he should have allowed, so the Contractor is unlikely to be compensated for the full value of dealing with the physical condition. If he should have allowed time and/or money within his tender for dealing with a physical condition then this must be considered in assessing a compensation event.

Note that, under Clause 60.2, in judging the physical conditions for a compensation event, the Contractor is assumed to have taken into account:

- the Site Information – this could include site investigations, borehole data, etc.
- publicly available information referred to in the Site Information – this could include reference to public records about the Site
- information obtainable from a visual inspection of the Site – note the use of the term "visual"; the Contractor is not assumed to have carried out an intrusive investigation of the Site
- other information which an experienced Contractor could reasonably be expected to have or to obtain – this is a fairly subjective criteria, but precludes the Contractor from relying solely on what is contained within the Site Information. Note the clause refers to "an experienced Contractor" not one who has a detailed local knowledge of the Site.

Under Clause 60.3, "if there is an ambiguity or inconsistency within the Site Information (including the information referred to in it), the Contractor is assumed to have taken into account the physical conditions more favourable to doing the work", this could be the cheaper, easier or quicker alternative.

This complies with the rule of "contra proferentem" (contra = against; proferens = the one bringing forth) in that where a term or part of a contract is ambiguous or inconsistent it is construed strictly against the party which imposes or relies on it.

It is critical that the Employer makes all information in his possession available to the Contractor. He cannot be selective, withholding information with a view of obtaining advantage. It is also important to note that if the Contractor encounters unforeseen physical conditions, which are often but not always, ground conditions, he may not necessarily be compensated for the cost and time effect of dealing with it, so he must consider carefully the wording of Clause 60.1(12) and 60.2 to prove his case.

It is also important to recognise that compensation to the Contractor is assessed as the difference between what the Contractor found, and what it would have been reasonable for him to have allowed in his tender, not simply the difference between what he found and what he did allow in his tender.

Engineering and Construction Short Contract

The Engineering and Construction Short Contract has exactly the same provisions but under Clause 60.1(9).

Professional Services Contract

There is no provision for unforeseen physical conditions within the Professional Services Contract.

Term Service Contract

There is no provision for unforeseen physical conditions within the Term Service Contract.

Question 6.2 How do the NEC3 contracts deal with delays and/ or costs incurred by exceptionally adverse weather conditions?

The NEC3 contracts deal with weather in different ways.

Engineering and Construction Contract

Under the Engineering and Construction Contract, we must consider Clause 60.1(13):

"A weather measurement is recorded:

- within a calendar month
- before the Completion Date for the whole of the works, and
- at the place stated in the Contract Data the value of which, by comparison with the weather data, is shown to occur on average less frequently than once in ten years. Only the difference between the weather measurement and the weather which the weather data show to occur on average less frequently than once in ten years is taken into account in assessing a compensation event."

With regard to weather related delays, most contracts use the words "exceptionally adverse weather" or "exceptionally adverse climatic conditions" leaving it to the parties to determine, and in many cases to argue, what is meant by the words "exceptionally adverse".

"Exceptionally" may be defined as "more than normal" whilst adverse may be defined as "unfavourable" in the sense that it impacted upon the Contractor's progress, which has to take into account the weather that would reasonably have been expected bearing in mind the location of the site, the time of year and what the Contractor was intending to do at the time the weather condition occurred.

The Engineering and Construction Contract provides a more objective approach by referring and comparing to weather data provided by an independent party such as a meteorological office, an airport or military base.

Only the difference between the weather measurement and the weather which the weather data show to occur less frequently than once in ten years is taken into account in assessing a compensation event, so the Contractor will not be compensated for the full impact of the weather event as weather likely to occur within a ten year period is the Contractor's risk, and it is assumed that the Contractor has already allowed for that in terms of price and programme. Contract Data Part 1 defines the place where the weather is to be recorded. Note that Clause 60.1(13) states "at the place stated in the Contract Data", so it is vital that the historic weather data and the current weather measurements are recorded at the same place. It is important that the place chosen will truly reflect the likely conditions which will be encountered at the Site.

It also lists weather measurements for each calendar month in respect of:

- the cumulative rainfall (mm)
- the number of days with rainfall more than 5 mm – whilst location and time of year is clearly a factor, from analysis of meteorological records, 5 mm rainfall in a day is a fairly low level
- the number of days in the month with minimum air temperature less than 0 degrees Celsius, and
- the number of days in the month with snow lying at a stated time GMT – there is no measure of how much snow, merely that it is lying, presumably on the ground at a stated time. There is also provision for adding other measurements which could include wind speed and other weather related data. This might apply at, say, coastal or mountain locations.

Note also that the weather measurement is recorded before the Completion Date for the whole of the works and at the place stated in the Contract Data.

It has to be said that, although the NEC drafters have recognised that weather and its effects should be more objectively measured and managed, there still remains some confusion as to when a weather-related compensation event exists and how to assess it. Ironically, it is even more important to assess the event correctly with Engineering and Construction Contract than with other contracts, as the Contractor is awarded time as well as money, whereas other contracts tend to only award time.

As a compensation event under Clause 60.1(13) occurs when a weather measurement is recorded within a calendar month, the value of which, by comparison with the weather data, is shown to occur on average less frequently than once in ten years, there is a "trigger point" in the month at which the weather, for example the cumulative rainfall exceeds the "once in ten years" test.

Note that the compensation event occurs once the trigger point has been exceeded, not once it has an effect on the progress of the works. The weather up to the trigger point is at the Contractor's risk in terms of programme and cost and is deemed to have been included within the Contractor's tender, but once the trigger point is reached, there are two schools of thought as to how the compensation event operates.

1 If the trigger point is reached on, say, the 25th of the month, one must consider the whole month, but only the difference between the weather measurement and the weather which the weather data show to occur on average less frequently than once in ten years is taken into account in assessing a compensation event.
2 If the trigger point is reached on, say, the 25th of the month, only delays incurred by the weather after the 25th are compensated for. In either case, one must consider the worst weather in ten years, not the average weather. Also, once a compensation event exists, it is incumbent upon the Contractor to submit a quotation showing any effect that the weather had on the Prices and/or any delay to Completion. There is no automatic entitlement to be paid for the weather merely because the trigger point was passed.

Engineering and Construction Short Contract

Under the Engineering and Construction Short Contract, we must consider Clause 60.1(10):

"The Contractor is prevented by weather from carrying out all work on the site for periods of time, each at least one full working day, which are in total more than one seventh of the total number of days between the starting date and the Completion Date. In assessing this event, only the working days which exceed this limit and on which work is prevented by no other cause are taken into account."

This is a departure from the statistical basis in the Engineering and Construction Contract. It is much simpler and easier to implement. It has the effect that the Contractor takes the risk of adverse weather affecting the works up to a certain threshold, i.e. one-seventh of the total number of days in the contract period. Beyond that threshold if the Contractor is prevented from carrying out *all* work on the site and for *complete* working days by the weather and *no other cause*, then it is a compensation event. Adequate records will need to be kept to establish such a compensation event.

The problem with the wording of this clause, as distinct from the more comprehensive wording within Clause 60.1(13) of the Engineering and Construction Contract, is that it does not take into account "normal" weather for that time of year at that location – e.g. it could be an exposed coastal site where the weather conditions are adverse for most of the time.

Professional Services Contract

Unlike the NEC3 Engineering and Construction Contract and Engineering and Construction Short Contract, the Professional Services Contract does not include a compensation event for delays and/or costs caused by adverse weather.

Whilst this may at first not seem to be an important issue for the Consultant – for example, in a drawing office designing a project rather than the Contractor on site building it – it can be a concern and a major risk for a Consultant carrying out other professional services which may be affected by the weather – for example, site inspections, surveys, etc.

Where external work is involved, it may be appropriate for the Employer to add a Z clause for weather-related risks similar to Clause 60.1(13) of the Engineering and Construction Contract.

Term Service Contract

There is no provision for delays and/or costs incurred by adverse weather conditions under the Term Service Contract.

Question 6.3 Do the NEC3 contracts include provision for "force majeure" events?

No, the NEC3 contracts do not specifically provide for "force majeure" events in the sense of describing them, and defining what is a force majeure event, though, if we consider the Engineering and Construction Contract, this includes Clause 60.1(19), which many could define as a "force majeure type of event".

The clause states:

"An event which:

- stops the Contractor completing the works, or
- stops the Contractor completing the works by the date shown on the Accepted Programme, and which
- neither Party could prevent
- an experienced contractor would have judged at the Contract Date to have such a small chance of occurring that it would have been unreasonable for him to have allowed for it, and
- is not one of the other compensation events stated in this contract."

Clause 60.1(19) is a new "no fault" clause introduced under NEC3, and dealing with an event neither Party could prevent, which stops the Contractor completing the works.

It is also tied in to Clause 19.1 "Prevention" following which, if the event occurs, the Project Manager instructs the Contractor how to deal with the event. There has been much confusion and debate regarding these clauses since NEC3 was first published in 2005.

The intention of the drafters appears to be that this would be a force majeure (superior force) clause, often referred to as an "Act of God" in other contracts and legal documents, i.e. significant events such as flooding, earthquakes, volcanic eruptions and the associated dust clouds and other natural disasters which prevent the contracting parties from fulfilling their obligations under the contract, the clause essentially freeing them, or giving some relief from their liabilities. Whilst this appears to be what the drafters had intended, on closer examination the application of this clause is far wider. Let's take each element in turn.

The event:

- stops the Contractor completing the works, or due to the event, the works were never completed
- stops the Contractor completing the works by the date shown on the Accepted Programme; due to the event, completion of the works was delayed
- could not be prevented by either Party; clearly, if either party could have prevented the event then they would or should have (the clause refers to preventing the event, rather than taking some mitigating action to lessen its effect)
- would have been judged, by an experienced contractor at the Contract Date, to have such a small chance of occurring that it would have been unreasonable for him to have allowed for it; at tender stage, the likelihood of the event was so low, the Contractor would not have allowed for it
- is not one of the other compensation events stated in this contract; if the matter could have been dealt with as one of the other compensation events then it should be. Clause 60.1(19) then clearly deals with the exceptional event.

Let's look at an example of where this clause would at least cause confusion.

Clause 60.1(13) has been deleted from the contract by means of a Z clause; one would then assume that the risk of unforeseen weather conditions is the Contractor's and he should have allowed for it. The site is affected by a severe storm that occurs less frequently than once in 50 years, which delays completion of the works by one week.

The Contractor states that:

- the weather stopped him completing the works by the date shown on the Accepted Programme; assume that the Contractor can prove that
- neither party could prevent the event; whilst one can take measures to lessen the impact of a storm, one cannot prevent the storm from occurring
- [an experienced Contractor states that] he would have judged at the Contract Date to have such a small chance of occurring that it would have been unreasonable to have allowed for it; the Contractor states that he could not have predicted or allowed for such a rare event and its consequences
- it is not one of the other compensation events.

If Clause 60.1(13) has been deleted from the contract by means of a Z clause, one could argue that it is no longer one of the other compensation events.

Could Clause 60.1(19) also extend to loss or damage occasioned by insurable risks such as fire, lightning, explosion, storm, flood, earthquake, terrorism?

Although both Schedules of Cost Components require that the cost of events for which the contract requires the Contractor to insure should be deducted from cost, the Contractor could still recover any delay to Completion through the compensation event.

If Clause 60.1(19) is intended as a form of force majeure provision, rather than including a fairly long-winded definition within the clause, why not just state "A force majeure event occurs", and allow the parties to interpret this internationally recognised term as they do in other contracts?

Question 6.4 How long does the Contractor on an NEC3 Engineering and Construction Contract have to notify the Project Manager of a compensation event?

Compensation events may either be notified by the Project Manager, or the Contractor can notify a compensation event to the Project Manager.

1. The Project Manager notifies the Contractor of the compensation event at the time of giving the instruction (Clause 61.1). He also instructs the Contractor to submit quotations, unless the event arises from a fault of the Contractor or quotations have already been submitted. The Contractor puts the instruction or changed decision into effect.
2. If the Project Manager instructs the Contractor to submit a quotation for a proposed instruction, the Contractor has only been instructed to submit a quotation, he does not carry out the work until instructed to do so.
3. The Contractor can notify a compensation event, but must do so within eight weeks of becoming aware of the event, otherwise he is not entitled to a change in the Prices, the Completion Date or a Key Date, unless the Project Manager should have notified the event to the Contractor but did not (Clause 61.3).

The intention of the clause is to compel the Contractor to notify compensation events promptly, otherwise any entitlement to additional time and money is lost. Note that the eight week rule requires the Contractor to notify the compensation event to the Project Manager within eight weeks of becoming aware of it, not just to have given early warning, or mentioned the possibility of a compensation event in a discussion with the Project Manager, or had his comments included in minutes of a meeting.

The clause therefore covers compensation events initiated by the Contractor, rather than the Project Manager, examples being the weather, unforeseen ground conditions, which the Project Manager may not be aware of unless the Contractor had notified it.

An example of the Project Manager failing to notify a compensation event when he should have would be if an instruction was issued by the Project Manager to the Contractor changing the Works Information, but at the time the Project Manager did not notify the compensation event, and the Contractor in turn did not notify either.

Even if the compensation event is not notified within eight weeks, the responsibility remains with the Project Manager as he gave the instruction and should have notified the event to the Contractor but did not.

Various opinions have been published about the enforceability and effectiveness of time bars in contracts such as NEC3, commentators particularly debating whether the clause is a condition precedent to the Contractor being able to recover time and money, and whether a party, the Employer, can benefit from its own breach of contract to the detriment of the injured party, the Contractor.

For example, if the Employer does not provide something which he is to provide by the date for providing it shown on the Accepted Programme, this is a valid compensation event under Clause 60.1(3), but can the Employer prevent the Contractor from receiving any remedy because the Contractor failed to notify the Project Manager within the eight week limit, and possibly, if Completion is delayed the Contractor have to pay delay damages?

The author, whilst not being a lawyer, is of the opinion that the parties are clear as to what the terms of their agreement are at the time the contract is formed. It is also clear what happens if the Contractor does not notify a compensation event within the time stated within Clause 61.3; he is protected against notifications which the Project Manager should have given but did not, and therefore the time bar must be effective and enforceable.

Clause 61.4 covers three possible outcomes to the Contractor's notification of a compensation event under Clause 61.3.

Negative reply: If the Project Manager responds by stating that an event notified by the Contractor:

- arises from a fault of the Contractor
- has not happened and is not expected to happen
- has no effect upon Defined Cost, Completion or meeting a Key Date, or
- is not one of the compensation events stated in the contract

he notifies the Contractor of his decision that the Prices, Completion Date and the Key Dates are not to be changed.

The Project Manager only needs to name one of these as his reason for refusing a compensation event.

Positive reply: If the Project Manager decides otherwise, he notifies the Contractor that it is a compensation event, and instructs him to submit quotations.

No reply: If the Project Manager does not reply within one week of the Contractor's notification, or a longer period to which the Contractor has agreed, then the Contractor may notify the Project Manager to that effect. If the Project Manager does not reply within two weeks of the Contractor's notification, then it is deemed to acceptance that the event is a compensation event and an instruction to submit quotations. It is a condition precedent upon the Contractor that the notification be given before deemed acceptance can occur.

Question 6.5 We are working on an Option A (priced contract with activity schedule) under the Engineering and Construction Contract and have submitted quotations for compensation events to the Project Manager, but he has not responded, and has said he will discuss them with us when we submit the Final Account. Can he do this?

First, there is no such thing as a Final Account under any of the NEC3 contracts, it is not a term that has ever been used!

The NEC3 contracts start with the Total of the Prices at the Contract Date and these prices are amended by compensation events which are assessed within the timescales of the contract.

There has always been concern amongst Contractors who were working under the 1st and 2nd editions of the Engineering and Construction Contract who queried what remedy was available if the Project Manager failed to reply to a quotation within the period required by the contract. The simple answer was that failure to reply to a communication within the period required by the contract was another compensation event under Clause 60.1(6), which does not remedy the problem with the initial compensation event.

The drafters of NEC3 considered the problem and introduced a new Clause 62.6, which, in seeking to provide a remedy to the Contractor for the Project Manager's failure, seems to favour the Project Manager.

If the Project Manager does not reply to a quotation within the time allowed, the Contractor may notify him to that effect. If the Contractor has submitted more than one quotation he states in that notification which one he proposes is to be accepted. If the Project Manager does not reply within two weeks of the notification, the notification from the Contractor is treated as acceptance of the Contractor's quotation by the Project Manager (Clause 62.6).

It is somewhat unbalanced that a pre-condition to deemed acceptance is that the Contractor must issue a notification giving the Project Manager a further but final two weeks in which to deliberate, yet if the Contractor fails to provide a quotation within the time allowed, the Project Manager's notification tells the Contractor that he will be making his own assessment, so there is no further time for the Contractor to respond.

In addition, Secondary Option W1 (Dispute Resolution) specifically allows the Employer to refer a dispute about a quotation for a compensation event which is treated as having been accepted, to be referred to the Adjudicator, so the Contractor still does not have final acceptance!

This is a curious inclusion when Clause 62.6 states that no reply to the Contractor's notification is treated as acceptance by the Project Manager. One would assume that an adjudicator whose role is to enforce the contract would have to find in favour of the Contractor.

Question 6.6 How should a Contractor under the NEC3 Engineering and Construction Contract prepare a quotation for a compensation event?

Quotations for compensation events are based on their effect on Defined Cost and time. This is different from most standard forms of contract where variations are valued using the rates and prices in the contract as a basis. The reason for this policy within the Engineering and Construction Contract is that no compensation event which is the subject of a quotation is due to the fault of the Contractor or relates to a matter which is at his risk under the contract. It is therefore appropriate to reimburse the Contractor his forecast additional costs or Defined additional costs arising from the compensation event.

Compensation events are assessed as the effect of the compensation event upon:

- the actual Defined Cost of work already done
- the forecast Defined Cost of the work yet to be done, and
- the resulting Fee.

Under Options A to D, if the Project Manager and Contractor agree, rates and lump sums may be used instead of Defined Cost. Such rates and prices do not have to be from the Activity Schedule (Options A and C) or Bill of Quantities (Options B and D), thereby allowing a mutually agreed fair rate or price to be agreed.

The Contractor must ensure that he includes within his quotation for cost and time which have a significant chance of occurring and are his risk (Clause 63.6). This should include for adverse weather conditions and physical conditions which would not be compensation events.

Changes to the Prices take the form of changes to the Activity Schedule (Options A and C) or changes to the Bill of Quantities (Options B or D).

Under Clause 63.3, a delay to the Completion Date is assessed as the length of time that, due to the compensation event, planned Completion is later than planned Completion as shown on the Accepted Programme.

A delay to a Key Date is assessed as the length of time that, due to the compensation event, the planned date when the Condition stated for a Key Date will be met is later than that shown on the Accepted Programme.

Clearly, this emphasises the need for an Accepted Programme to be in place in order that it can be used to make correct assessments.

No compensation event can result in a reduction in the time for carrying out the works, i.e. an earlier Completion Date. Only acceleration as agreed under Clause 36 can result in an earlier Completion Date.

Any time risk allowances for which the Contractor has allowed are preserved by this clause, as assessment of the compensation event is based on entitlement rather than need. Allowances for risk must be included in forecasts of Defined Cost and Completion in the same way that the Contractor allows for risks when pricing his tender. Float within the Accepted Programme is, however, available to mitigate or avoid any consequential delay to the Completion Date.

Delay to Completion

First, the Contractor considers whether there will be any delay to planned Completion. The Contractor should provide details to demonstrate to the Project Manager the basis on which the delay has been calculated, this will probably require him to issue a programme, or an extract from the programme with his quotation.

Changes to the Prices

First, it must be remembered that the change to the Prices is not assessed by using the Contractor's rates from the activity schedule or bill of quantities. It must be assessed as the effect of the compensation event upon the Defined Cost of the work and the resulting fee.

Note that, if the Project Manager and Contractor agree, rates and lump sums may be used to assess a compensation event instead of Defined Cost. This may be done either by using rates and lump sums from the activity schedule or bill of quantities, or by simply mutually agreeing other rates and lump sums.

Question 6.7 We are Contractors carrying out a refurbishment contract under NEC3 Engineering and Construction Contract Option A (priced contract with activity schedule) and have had several individual compensation events each with time delays of less than one day. However, the total cumulative effect of all these compensation events will cause a delay to the Completion Date of approximately nine days. How do we recover these delays and their associated costs under the contract?

Under the NEC3 Engineering and Construction Contract (Clause 62.2), the Contractor must assess and include within his quotation for a compensation event for any delay to the Completion Date and Key Dates, and submit details of his assessment within his quotation. In assessing any delay to the Completion Date the delay is the length of time that due to the compensation event planned Completion is later than planned Completion as shown in the Accepted Programme.

This is fine where the Contractor forecasts that a compensation event will impact on the Completion Date by, say, two weeks; the delay and any associated effect on Defined Cost plus Fee will then be included within the quotation.

The problem is that, often with refurbishment projects, a compensation event may arise as a result of a Project Manager's instruction to carry out a minor piece of work – for example, to remove and replace the skirtings to part of a room instead of retaining and repainting the old skirtings.

In this case, the Contractor will assess the effect on his Defined Cost (plus Fee percentage) including removing the old skirtings, making good the wall, replacing with new skirtings and painting, but although it may in theory add a couple of hours to the programme and in turn the completion of the project, the Contractor would probably not include anything for delays within his quotation, and in fact that compensation event on its own would not cause any delay to the Completion Date anyway, so he could not include for the delay and the associated costs.

However, if this happens on several occasions, which again could be likely on a refurbishment project, then the cumulative delay could become apparent.

Whilst the author is firmly of the belief that parties should fully comply with the contract, sometimes you have to apply a practical and common sense solution to its compliance.

With situations like this it is probably best to mutually agree between the Contractor and the Project Manager that compensation events may be "batched together" maybe in terms of "changes to scope of work for skirtings" or just collected on a monthly basis to give a collective and cumulative effect on Defined Cost plus Fee, and the Completion Date.

In order to maintain some control, the Project Manager should state assumptions about the event in his instruction to the Contractor to submit each quotation – e.g. "do not include for any delay to the Completion Date" – then notify a correction to an assumption under Clause 61.6 and then deal with it under Clause 60.1(17) as a compensation event.

Question 6.8 We have a Consultant carrying out survey work under Option A (priced contract with activity schedule) of the NEC3 Professional Services Contract. The survey has been substantially delayed due to extremely bad weather conditions. Is such a delay the Consultant's own risk?

As has been stated previously, the Professional Services Contract has no equivalent of the Engineering and Construction Contract Clause 60.1(13) which gives a compensation event for a weather related event, so one would then assume that the risk of unforeseen weather conditions in this case is the Consultant's and he should have allowed for it.

But let us consider Clause 60.1(11) which could, at least in theory change this.

Clause 60.1(11) states:

"An event which:

- stops the Consultant completing the services or
- stops the Consultant completing the services by the date shown on the Accepted Programme,

and which:

- neither Party could prevent
- an experienced consultant would have judged at the Contract Date to have such a small chance of occurring that it would have been unreasonable for him to have allowed for it, and

is not one of the other compensation events stated in this contract."

The Consultant's progress with an external survey is affected by a very severe storm which occurs less frequently than once in 50 years, which delays completion of the works by 1 week.

The Consultant states that:

- the weather stopped him completing the survey by the date shown on the Accepted Programme; assume that the Consultant can prove that
- neither party could prevent the event; whilst one can take measures to lessen the impact of a storm, one cannot prevent the storm from occurring
- [an experienced Consultant states that] he would have judged at the Contract Date to have such a small chance of occurring that it would have been unreasonable to have allowed for it; the Consultant states that he could not have predicted or allowed for such a rare event and its consequences
- it is not one of the other compensation events; there is no weather-related compensation event in the contract, but has Clause 60.1(11) now created one?

Question 6.9 We have a compensation event where, in discussion with the Contractor during a risk reduction meeting he has stated that it is extremely difficult to price one element of the work, and we agree with him. The Contractor has suggested including a Provisional Sum within his quotation, which he can adjust later. Can he do this?

First, it is essential to note that the NEC3 contracts do *not* provide for Provisional Sums as used in other contracts where there are elements of work which are not designed or cannot be sufficiently defined at tender stage, and therefore a sum of money is included by the Employer in the Bill of Quantities or other pricing document to cover the item.

When the Contractor is instructed to submit a quotation, there may be a part of the quotation which is too uncertain to be forecast reasonably. In this case the Project Manager should state what the Contractor should assume, which could be a cost and/or time effect, this is called an assumption.

Subsequently, when the effects are known or it is possible to forecast reasonably the Project Manager notifies a correction to the assumption and it is dealt with as a correction under Clause 60.1(17) (see Example under Clause 60.1(17) earlier).

Note, that the Project Manager must state the assumption; if the Contractor makes assumptions when pricing the compensation event, then they are not corrected.

There is provision for assumptions within:

- NEC3 Engineering and Construction Contract
- NEC3 Engineering and Construction Short Contract.

Question 6.10 How should a Project Manager on an NEC3 Engineering and Construction Contract make his own assessment of a compensation event?

Under Clause 64.1–4, the Project Manager may, for the following reasons, assess a compensation event:

- if the Contractor has not submitted a required quotation and details of his assessment within the time allowed
- if the Project Manager decides that the Contractor has not assessed the compensation event correctly in a quotation and he does not instruct the Contractor to submit a revised quotation

- if, when the Contractor submits quotations for a compensation event, the Contractor has not submitted a programme which the contract requires him to submit, or
- if, when the Contractor submits quotations for a compensation event, the Project Manager has not accepted the Contractor's latest programme for one of the reasons stated in the Contract.

These are all derived from some failure of the Contractor either to submit a quotation, to assess the compensation event correctly or to submit an acceptable programme.

If the Project Manager makes his own assessment he should put himself in the position of the Contractor, giving a properly reasoned assessment of the effect of the compensation event, detailing the basis of his calculations and providing the Contractor with details of that assessment.

A Project Manager's assessment is not simply the quotation returned to the Contractor with "red pen" reductions down to a figure the Project Manager is prepared to accept or to get the Employer to pay!

It is also important to recognise that the Project Manager is not sending his assessment to the Contractor for his acceptance; it is the final decision under the contract, the Contractor's only remedy being adjudication.

Under Clause 64.4, if the Project Manager does not assess a compensation event within the time allowed, the Contractor may notify him to that effect. If the Project Manager does not reply to the notification within two weeks, the Contractor's notification is treated as acceptance of the quotation by the Project Manager. The only remedy the Project Manager has, if he later finds the quotation is not acceptable, is to refer it to adjudication.

However, it must be remembered that the role of the Adjudicator and the adjudication process is to enforce the contract; therefore, if the Contractor's quotation has been submitted in accordance with the contract, then the Employer will probably be unsuccessful in the adjudication.

Question 6.11 The Project Manager on an NEC3 Engineering and Construction Contract has instructed the omission of a part of the works. Can the Contractor claim for loss of profit on the omitted works?

When the Project Manager gives an instruction changing the Works Information by omitting work, the method of changing the Prices and dealing with the Completion Date is often misunderstood.

The tendency is for parties to simply delete or even to just ignore the item in the pricing document (Activity Schedule or Bill of Quantities). For example, in an Option A contract the activity in the Activity Schedule which represents the omitted work is simply reduced or not paid, or in an Option B contract the quantity in the Bill of Quantities is reduced, or if a single item in the bill, the item is just not paid.

Admittedly, this is usually the correct way to price change in a non-NEC contract, but in an NEC contract, both assumptions are incorrect. Omitted work due to a change to the Works Information is assessed not by omitting or remeasuring the item in the relevant pricing document, but by forecasting the Defined Cost of that omitted work and adding the Fee percentage. The resulting amount is then adjusted against the value in the pricing document.

This principle is clearly stated in Clause 63.1 of the contract – i.e. the changes to the Prices are based on:

- the actual Defined Cost of work already done
- the forecast Defined Cost of the work yet to be done, and
- the resulting Fee.

It must be remembered that the principle with compensation events in terms of their financial value is that neither party gains or loses as a result of the compensation event; the Contractor is compensated so that he is in the same financial position after the event as he would have been before the event.

If a Contractor has included low rates in his tender for work which is subsequently omitted, it is quite probable that the application of forecast defined Cost plus Fee will give rise to a negative value; this is testament to the fact that the Contractor would have made a loss if the work had not been omitted. Similarly if high rates exist, the Contractor will retain the margin he would have made if he had carried out the work.

The Contractor cannot directly claim for loss of profit, but he will retain the profit that he would have made if the omission had not occurred.

The same principle would apply with all the Main Options. In the example, if it was an Option B contract and the boundary fence was an item in the Bill of Quantities, then the Bill of Quantities would again be changed in the same way, and again the Contractor retains the profit (or loss) he would have made if the boundary fence had not been omitted.

Example

In an Option A contract, the Activity Schedule includes an activity entitled "boundary fence", the activity having a price of £26,000.

The Project Manager gives an instruction to omit the boundary fence; this is a change to the Works Information and therefore a compensation event under Clause 60.1(1).

The tendency is to assume that the Contractor is simply not paid for that activity, as it will not be carried out. However, that would be incorrect.

Let us assume that the Contractor has already placed an order in the sum of £20,100 with a fencing Subcontractor, and can therefore prove what his costs would have been had the work not been omitted and the Fee percentage is 8 per cent.

The Forecast Defined Cost is then the value of the subcontract order.

Subcontract Order to supply and install fencing= £20,100.00

+ 8% (£1,608.00)

Change to the Prices = – £21,708.00

The amount of the Contractor's quotation, assuming it has been accepted by the Project Manager, is then used to change the Prices, so either a new activity is then inserted to the value of –£21,708 or the activity priced at £26,000 is omitted and replaced with an activity priced at £26,000 – £21,708 = £4,292.

The Contractor in this case retains the profit he would have made if the boundary fence had not been omitted. Clearly, if the change to the Prices had been more than £26,000 the Contractor would retain the loss.

Question 6.12 We have an Engineering and Construction Contract Option B (priced contract with bill of quantities), and a description of a product in the Bill of Quantities states "or similar approved". If the Contractor then proposes a cheaper alternative (which is approved), does this saving get administered as a negative compensation event?

The phrase "or similar approved" must be stated in the Works Information, not the bill of quantities, as it is the Works Information that basically tells the Contractor what to do.

Assuming this is the case then, providing there has been such an approval, the Contractor has complied with the Works Information. There is no instruction needed to change it and there is no compensation event, negative or otherwise. The Works Information was written to allow this, providing it was approved, and that is what has happened. The financial effect of that depends on what Main Option you are using.

In Main Options A (priced contract with activity schedule) and B (priced contract with bill of quantities) the Contractor keeps all the benefit (or possibly the burden).

However, in Main Options C and D (target contracts with bill of quantities and activity schedule respectively) the benefit (or burden) is shared.

Note also that acceptance rather than approval is generally the language used in NEC contracts.

Question 6.13 We would like to include value engineering within the NEC3 contracts and to reward the Contractor if he proposes a change which can save the Employer money. Is there any provision for value engineering within the NEC3 contracts?

Many users of the NEC3 contracts do not realise that there is already provision for value engineering within the contracts, for example within the Engineering and Construction Contract Option C (Target Contract with Activity Schedule) and Option D (Target Contract with Bill of Quantities) there is Clause 63.11.

Under these options, if the Employer changes the Works Information – perhaps because his design team have carried out a value engineering exercise to generate savings – and he wishes to incorporate their findings within the Contract, the Project Manager gives an instruction changing the Works Information and it is a compensation event (Clause 60.1(1)); the Contractor then submits a quotation for the saving and the Prices are reduced (the target is lowered) accordingly, then the cost savings will be accrued through the Defined Cost paid to the Contractor.

However, the Contractor may have carried out a value engineering exercise and submitted proposals to the Project Manager for his (and the Employer's) consideration. If the Project Manager then accepts the proposals and the Works Information is changed accordingly then the first bullet of Clause 63.11 is applicable, i.e.:

If the effect of a compensation event is to reduce the total Defined Cost and the event is:

- a change to the Works Information, other than a change to the Works Information provided by the Employer which the Contractor proposed and the Project Manager has accepted …

the Prices are reduced.

This clause has to be read a couple of times as it seems quite convoluted, but what it means is that the Prices are reduced (the target is lowered) *unless* there is a change to the Works Information provided by the Employer which the Contractor proposed and the Project Manager has accepted, in which case the prices are *not* reduced and the cost savings accrued through the Defined Cost paid to the Contractor will lead to a sharing of the benefit between the Employer and the Contractor.

Question 6.14 What does the term "implemented" mean within the NEC3 contracts when referring to compensation events?

The word "implemented" has a specific meaning under NEC3 contracts.

Under Clause 65.1 of the NEC3 Engineering and Construction Contract for example, the Project Manager implements each compensation event by notifying the Contractor of his acceptance of the Contractor's quotation, notifying the Contractor of his own assessment, or a Contractor's quotation may be treated as accepted subject to Clause 64.4.

Under Clause 65.2, the assessment of a compensation event is not revised if a forecast upon which it is based is shown by later recorded information to have been wrong.

If the subsequent records of resources on work actually carried out show that achieved Defined Cost and timing are different from the forecasts included in the accepted quotation or in the Project Manager's assessment, the assessment is not changed. This gives the implementation of the compensation event *finality*.

There is a mistaken belief that Clause 60.1(8) – "the Project Manager or the Supervisor changes a decision which he has previously communicated to the Contractor" – allows the Project Manager to "undo" his assessment of a compensation event – i.e., to change his mind at a later date. This is incorrect; the word "decision", like the word "implementing", has a specific meaning under NEC3 contracts.

The contract refers to "decisions" under the following clauses:

- Clause 11.2(25) – The Project Manager decides Disallowed Cost.
- Clause 30.2 – The Project Manager decides the date of Completion.
- Clause 50.1 – The first assessment date is decided by the Project Manager.
- Clause 61.4 – The Project Manager decides whether an event notified by the Contractor is a compensation event.
- Clause 61.5 – The Project Manager decides that the Contractor did not give an early warning of the event which an experienced Contractor could have given.
- Clause 61.6 – The Project Manager decides that the effects of a compensation event are too uncertain to be forecast reasonably.
- Clause 63.5 – The Project Manager has notified the Contractor of his decision that the Contractor did not give an early warning of the event which an experienced Contractor would have given.
- Clause 64.1 – The Project Manager assesses a compensation event if he decides that the Contractor has not assessed the compensation event correctly.

The Project Manager assessing and implementing a compensation event is not the same as giving a decision, it must be stressed that the Project Manager does not have the authority to change his assessment once it has been implemented. The Project Manager's assessment is final and the only recourse would be adjudication between the Parties i.e. the Employer and the Contractor.

Question 6.15 We understand that the NEC3 contracts do not have a Final Account. Is this true and, if so, what is the process of finalising all matters under the contract?

It is correct that the NEC3 Contract does not have the equivalent of a Final Account, Final Certificate or Final Statement, which is found in most other contracts, certifying that the contract has fully and finally been complied with, and that issues regarding the work itself, defects, changes to the works (termed "variations" in most other contracts), and payments have all been dealt with.

Whilst some may say that, because of the compensation event provisions dealing with changes as the project progresses, a compensation event is not notified after the defects date so there is finality in that respect, and the issue of the Defects Certificate confirming correction of Defects, there is no need for a Final Certificate.

However, some questions could still remain as to whether all matters are final under the contract for example, under Option E (cost reimbursable contract) when is it too late for the Contractor to submit a cost for reimbursement?

Most Employers will tend to issue a form of Final Certificate or Final Statement for signature by the Employer (and/or the Project Manager) and the Contractor confirming that nothing remains outstanding under the contract.

Obviously any issues of latent defects and limitation of action, etc., will be dealt with under the relevant law of the contract.

Chapter 7

Title

Title to Plant and Materials, objects and materials within the Site

Question 7.1 During excavations for the foundations to a new building as part of a project to build a new school using the NEC3 Engineering and Construction Contract Option A (priced contract with activity schedule), the Contractor has discovered some archaeological remains. How should this discovery be dealt with under the contract?

If the Contractor discovers any object of value or of other interest, first it is important to recognise that he has no title to it, so he has no right of ownership of whatever he finds as the Employer has title to materials from excavation or demolition unless the Works Information states otherwise (Clause 73.2).

It is important in this respect that the Works Information clearly details any title the Contractor has to materials arising from excavation or demolition as the Contractor needs to know, and will allow within his tender for the options of selling, recycling or removing those materials. Also, if the Employer has title to the materials, he must clearly state what the Contractor is to do with those materials – for example, hand them to the Employer on Site, transport them to another location, etc.

In this particular case, dealing with archaeological remains – especially where human remains are also found – is a very specialist and prescribed process as there will normally be specific regulations, and also dependant on where the Site is, statutory requirements, regarding dealing with the discovery, allowing other third parties to examine what has been found before carrying out any further work, whether they should then be preserved in situ or removed from Site, and who has title to what is discovered.

This can also have a severe impact on the programme and the cost of the project, so it is vital that all possible Site and other investigations are carried out at the time of preparing the tender documents so that the Employer, the Contractor, Project Manager and any other parties are as aware as they can be of the likelihood of finding such remains and how to deal with them.

Obviously, in the case of a school, which may have to be completed in time for the start of a new academic term, this could be a serious issue, though on a positive note, the find could provide a great opportunity for the schoolchildren to be made aware of the importance of protecting our archaeological heritage!

Going back to the contract, under Clause 73.1 of the NEC3 Engineering and Construction Contract, the Contractor must notify the Project Manager upon finding anything as described in the question, who then instructs the Contractor how to deal with it. It is vital that the Contractor only takes instructions from the Project Manager in this respect, although in the case of archaeological remains, other parties such as local authorities, museums and educational establishments may have an interest in the find and how it is dealt with.

Initially, an early warning notice should be issued by the Contractor (or the Project Manager) allowing the parties to share their thoughts and opinions on the possible impact upon Price and Completion Date and probably a risk reduction meeting should be held to consider proposals, seek solutions and decide actions.

This is a compensation event under Clause 60.1(7), so it is important that the Contractor not only notifies the Project Manager of a compensation event, if the Project Manager has not already notified, but also keeps detailed records of any stoppages, disruption and any changes to working methods as a result of the find, as he will need to substantiate the effect on Defined Cost, and also on any delay to the Completion Date.

Chapter 8

Indemnity, insurance and liability

Insurance requirements, claims, etc.

Question 8.1 What are the insurance requirements within the NEC3 contracts?

Let us take the NEC3 Engineering and Construction Contract as our first example.

Clause 80.1 lists the Employer's risks. However, these are the Employer's risks only in terms of loss, wear or damage, and are not an exhaustive list of all the risks the Employer bears under the contract.

There are many other risks such as actions or inactions of the Project Manager or Supervisor, unforeseen physical conditions and weather, which are covered elsewhere as compensation events.

There are six main categories of Employer's risk within this clause:

1 Risks relating to the Employer's use or occupation of the Site, negligence, breach of statutory duty or interference with any legal right, or a fault in his design. In respect of design, the Employer should cover the risk himself through professional indemnity insurance, or if he uses external consultants for the design, he should ensure that they hold such insurance. This insurance should be held for the full period of liability, which will normally be several years after completion of the works.
2 Risks relating to items supplied by the Employer to the Contractor until the Contractor has received them, or up to the point of handover to the Contractor. The Employer should ensure that he has adequate insurance in this respect, or again ensure that Others who supply items have such insurance.
3 Risks relating to loss or damage to the works, Plant and Materials, caused by matters outside the control of the Parties.

4 Risks arising once the Employer has taken over completed work, except a defect that existed at take-over, an event which was not an Employer's risk, or due to the activities of the Contractor on Site after take-over.
5 Risks relating to loss or wear or damage to parts of the works taken over by the Employer, and any Equipment, Plant and Materials retained on Site after termination.
6 Any other risks referred to in the Contract Data. These should be stated in Contract Data Part 1. Any risks not carried by the Employer are carried by the Contractor from the starting date to the issue of the Defects Certificate or a Termination Certificate. The Contractor is also obliged until the Defects Certificate has been issued and unless instructed by the Project Manager to promptly replace loss of and repair any damage to the works, Plant or Materials. Each party indemnifies the other against any claims or proceedings which are at their risk, although if a party partly contributed to the event.

The Insurance Table in the contract itemises the insurances that the Contractor has to effect together with the minimum amount of cover or minimum limit of indemnity. The default is that the Contractor provides, maintains and pays for the insurances, unless the Employer states otherwise in Contract Data Part 1.

Whether the Contractor or the Employer takes out the insurances, they are always effected as a joint names policy effective until the Defects Certificate or a termination certificate has been issued.

The Insurance Table

The four insurances listed within the Insurance Table are as follows.

(i) Loss or damage to the works, Plant or Materials

This item in the Insurance Table covers loss or damage to the works, including any Plant or Materials provided by the Employer. Some other contracts provide an option for either the Contractor or the Employer to provide this insurance, particularly where the works relate to an existing structure or building, where the Employer may already have an existing insurance policy.

These insurances are effected as a Joint Names Policy effective until the Defects Certificate or a Termination Certificate is issued and includes full reinstatement, replacement or repair (the amount stated should also include an amount or percentage to cover Professional Fees). Insurance of the Works will normally be covered by the Contractor's All Risks (CAR) policy.

The minimum cover for all insurances is stated in the Insurance Table. However, the Contractor is liable for whatever the amount of any claim, therefore he must consider the minimum value in the Insurance Table purely as a guide.

(ii) Loss or damage to Equipment

Again, the CAR policy should cover this. The reference to "replacement cost" means the cost of replacement with Equipment of similar age and condition rather than "new for old".

(iii) Loss or damage to property (except the works, plant or materials) or bodily injury or death not an employee of the Contractor

This requires the Contractor to indemnify the Employer against any loss, expense, claim, etc., in respect of any personal injury or death caused by the carrying out of the work, other than their own employees. This includes the liability towards members of the public who may be affected by the construction work, although they have no part in it. In the case where a party makes a claim directly against the Employer due to a death or injury the Contractor should either take on that claim, or alternatively, the Employer can sue the Contractor to recover any monies.

The Insurance Table states the minimum amount of cover or minimum limit of indemnity. The Contractor could be liable for whatever the amount of any Claim; therefore, he must consider the minimum value in the Insurance Table purely as a guide.

The Contractor must be able to prove, if requested by the Employer, that he has the required cover and his premiums are up to date, as if he does not, the Employer may take out Insurances himself and deduct the premiums from the Contractor.

The Contractor also used to have to prove that his Subcontractors had the same cover. However, this has now been amended, as Contractors can either ensure that the Subcontractors have the cover or can include it under their own cover on behalf of their Subcontractors.

(iv) Death or bodily injury to employees of the Contractor

This covers the liabilities as an employer of people to insure against injury or death caused to people whilst carrying out their work, which is a legal obligation in most countries.

The Contractor must be able to prove, if requested by the Employer, that he has the required cover and his premiums are up to date, as if he does not, the Employer may take out Insurances himself and deduct the premiums from the Contractor.

The Contractor also used to have to prove that his Subcontractors had the same cover. However, this has now been amended, as Contractors can either ensure that the Subcontractors have the cover or can include it under their own cover on behalf of their Subcontractors.

Again, the Insurance Table states the minimum amount of cover or minimum limit of indemnity, the Contractor having to insure for whatever the amount of any claim.

Engineering and Construction Short Contract

The Short Contract has similar insurance requirements including a similar Insurance Table, but with no express requirements in respect of the Contractor having to submit certificates showing that the relevant insurances are in place, if the Contractor does not insure, or for insurances to be provided by the Employer.

In addition, Clause 80.1 provides for a possible limit of the Contractor's liability to the Employer for loss or damage to his property if stated in the Contract Data.

Careful thought needs to be given to the limit specified as much will depend on the value, likelihood and consequence of damage, and also proximity of the Employer's property which may be at risk.

Professional Services Contract

The Professional Services Contract is again similar to the Engineering and Construction Short Contract, with a clause dealing with limitation of the Consultant's total liability to the Employer, again if stated in the Contract Data, but the first Insurance covers claims made against the Consultant arising out of his failure to use the skill and care normally used by professionals, etc. This is Professional Indemnity Insurance (PI) rather than the provisions in the Construction Contracts dealing with loss or damage to the works, Plant and Materials.

Term Service Contract

The Term Service Contract is similar to the Engineering and Construction Contract, without the limitation clause.

Supply Contract

The Supply Contract is similar to the Engineering and Construction Short Contract, but the first insurance covers loss or damage to goods, plant and materials rather than the provisions in the Construction Contracts dealing with loss or damage to the works, Plant and Materials.

Question 8.2 Can we include a requirement in an NEC3 contract for a Contractor or a Consultant to have Professional Indemnity (PI) insurance?

Let us first consider the subject of Professional Indemnity (PI) insurance.

If a Contractor or Consultant providing a service to an Employer makes a mistake, is found to be negligent, or gives inaccurate advice, then he will be liable to the Employer in event that the Employer incurs a loss as a result. This loss can be very significant where the design has to be corrected, parts of the structure have to be taken down and reinstated, a facility has to be closed down whilst the remedial measures take place, and there are also legal costs.

Professional Indemnity claims can arise where there is negligence, misrepresentation or inaccurate advice which does not give rise to bodily injury, property damage or personal injury, but does give rise to some financial loss. Additional coverage for breach of warranty, intellectual property, personal injury, security and cost of contract can be added. In that event, although the Employer claims against the Contractor or Consultant rather than from the insurers, PI Insurance protects the Contractor or Consultant against claims for loss or damage made by a client or third party.

If we first consider the NEC3 Professional Services Contract, within Clause 81.1, the Insurance Table includes the "liability of the Consultant for claims made against him arising out of his failure to use the skill and care normally used by professionals providing services similar to the services" (i.e. the services defined under the specific contract); this is Professional Indemnity (PI) insurance.

The minimum amount of cover is then stated within Contract Data Part 1 together with the period following Completion of the whole of the services which the insurance must be maintained for, unless there is an earlier termination.

In respect of the period for which the insurance is maintained, it is very important to mention that Professional Indemnity insurance policies are based on a "claims made" basis, meaning that the policy only covers claims made during the policy period when the policy is "live", so claims which may relate to events occurring before the coverage was active may not be covered.

However, these policies may have a retroactive date which can operate to provide cover for claims made during the policy period but which relate to an incident after the retroactive date.

The Project Manager, on behalf of the Employer, should ensure that the Contractor has taken out and maintained the required insurances for the full period of his liability. Claims which may relate to incidents occurring before the policy was active may not be covered, although a policy may sometimes have a retroactive date.

By default the NEC3 construction contracts where a party is the Contractor – i.e., the Engineering and Construction Contracts and the Term Service Contracts – do not specifically require the Contractor to have Professional Indemnity (PI) insurance, but Contract Data Part 1 provides for the Employer to insert a requirement for the Contractor to provide any additional insurances, which can include Professional Indemnity (PI) insurance.

Particularly where the Contractor is designing parts of the works, it would certainly be advisable for the Employer to include a requirement for the Contractor to provide Professional Indemnity (PI) Insurance, as although any Consultant employed by the Contractor may have relevant Professional Indemnity insurance, the liability to the Employer will lie with the Contractor, not with the Consultant he employs.

Chapter 9

Termination provisions

Reasons, procedures and amounts due

Question 9.1 We are Contractors on an NEC3 Engineering and Construction Contract Option A (priced contract with activity schedule), and we wish to terminate our employment under the contract due to non-payment by the Employer. Can we do this, and if so how?

The first thing to say is that termination should never be a decision taken lightly by any party under any contract, it really should be an action of last resort when the parties are not able to amicably resolve issues between themselves.

"Termination" is a word most commonly used in the context of construction contracts to refer to the ending of the Contractor's employment. The parties have a common law right to bring the contract to an end in certain circumstances, but most standard forms give the parties additional and express rights to terminate upon the happening of specified events.

Some contracts refer to termination of the contract, whilst others refer to the termination of the Contractor's employment under the contract. In practice, it makes little difference, and most contracts make express provision for what is to happen after termination.

Also, termination is an entitlement not an obligation, though clearly, if one of the parties has become insolvent, then there is no choice but to terminate the contract.

It is also important to note that neither party should attempt to terminate employment under the contract unless they are sure that the provision is available within the contract and if it is, then they must ensure that they strictly comply with the wording of the contract. If termination is held to be wrongful, it is usually a repudiation of the contract.

It should be noted that most contracts require a notice to be given by one party to the other unless the breach is due to insolvency. The offending party then has a period of time in which to remedy the breach, failing which termination can take place.

Termination, including reasons, termination procedures and payment, is covered within the Engineering and Construction Contract by Clauses 90.1 to 93.2 with further references within the Main Option clauses.

Within the Engineering and Construction Contract, there are 21 reasons listed for either party to terminate:

- Reasons R1 to R10 provide for either party to terminate due to the insolvency of the other. This includes bankruptcy, appointment of receivers, winding-up orders and administration orders dependent on whether the party is an individual, a company or a partnership.
- Reasons R11 to R13 provide for the Employer to terminate if the Project Manager has notified the Contractor that he has defaulted and the Contractor has not remedied the default within four weeks of the notification in respect of the Contractor substantially failing to comply with his obligations, not providing a bond or guarantee, or appointing a Subcontractor for substantial work before the Project Manager has accepted the Subcontractor. It is quite subjective as to what the words "substantially fail[ing] to comply" and "substantial work" mean within these provisions. One would assume that it is not intended to relate to minor breaches, but having said that, the Contractor could "substantially fail to comply" with a minor obligation.
- Reasons R14 to R15 provide for the Employer to terminate if the Project Manager has notified that the Contractor has defaulted and the Contractor has not stopped defaulting within four weeks of the notification in respect of substantially hindering the Employer or Others or substantially broke a health or safety regulation and not stopped defaulting within the four weeks of the notification. Again, the use of the word "substantially" is quite subjective. How would one substantially hinder someone, or substantially break a health or safety regulation? Is not a health or safety regulation either "broken" or "not broken"?
- Reason R16 provides for the Contractor to terminate if the Employer has not paid an amount certified by the Project Manager within 11 weeks of the date of the certificate.
- Reason R17 provides for either Party to terminate if the Parties have been released under the law from further performance of the

whole contract. Note the reference to "the whole contract", not just a part.

- Reasons R18 to R20 provide for the Project Manager having instructed the Contractor to stop or not restart any substantial work or all work and an instruction allowing the work to restart or start has not been given within 13 weeks. Either Party may terminate if the instruction was due to a default by the other, or if the instruction was due to any other reason.
- Reason R21 provides for the Employer to terminate if an event occurs which stops the Contractor completing the works, or stops the Contractor completing the works by the date shown on the Accepted Programme and is forecast to delay Completion by more than 13 weeks, and which neither Party could prevent and an experienced Contractor would have judged at the Contract Date to have such a small chance of occurring that it would have been unreasonable for him to have allowed for it.

Note that the Contractor can only terminate for one of the above reasons, though the Employer can terminate for any reason. Whilst this may seem inequitable, if the Employer terminates for a reason not stated in the Termination Table he will have to pay the Contractor for work carried out up to the termination, other costs incurred in the expectation of completing the whole of the works – for example, orders placed and other commitments made – the forecast Defined Cost of removing Equipment, and, dependent on the Main Option used, the direct fee percentage applied to the difference between the original total of the Prices and the Price for Work Done to Date, essentially a loss of profit/overheads provision.

Clearly, the earlier the Employer exercises this right of termination, the greater this amount. With all the termination reasons, the party wishing to terminate notifies the Project Manager and the other Party giving details of his reasons for terminating. The Project Manager then issues a termination certificate to both parties if he is satisfied that the reason for termination is valid under the contract. It is perhaps curious that the Project Manager, who acts for the Employer, makes the decision as to whether the termination is valid?

So if the Contractor on an NEC3 Engineering and Construction Contract Option A wishes to terminate their obligation to Provide the Works under the contract due to non-payment by the Employer, the first action is to check whether it is one of the reasons under the contract.

In this case it is Reason R16 (R16), "the Contractor may terminate if the Employer has not paid an amount due under the contract within 11 weeks of the date that it should have been paid", so clearly the 11 weeks is a critical factor in deciding whether the Contractor can terminate under the Contract.

So let us consider exactly how termination can be effected in the case of this Contractor.

First, the Contractor must notify the Project Manager and the Employer that he wishes to terminate and give details of his reasons for terminating. If the reason complies with the Contract, then the Project Manager issues a termination certificate to the Contractor and to the Employer.

After the termination certificate has been issued, the Contractor does no further work necessary to Provide the Works.

The procedures of termination are then implemented immediately upon issue of the termination certificate.

Procedures on termination

The Contractor leaves the Working Areas and removes his Equipment.

The Employer may complete the works, either himself or using another Contractor, and may use any Plant and Materials to which he has title.

Payment on termination

Within 13 weeks of termination, the Project Manager must certify a final payment to the Contractor, which is the Project Manager's assessment of the net amount due on termination, and payment is then made by the Employer to the Contractor within three weeks of the Project Manager's certificate.

The Contractor is entitled to be paid:

- an amount due as for normal payments, in this case the Contractor has been employed under Option A of the Engineering and Contraction Contract
- the Defined Cost for Plant and Materials within the Working Areas, or to which the Employer has title and of which the Contractor has to accept delivery
- other Defined Cost reasonably occurred by the Contractor in completing the works

- any amounts retained by the Employer, and
- a deduction of any un-repaid balance of an advance payment

also,

- the forecast Defined Cost of removing the Equipment

also,

- for Options A, B, C and D, the *direct fee percentage* applied to any excess of the total of the Prices at the Contract Date (referred to in other contracts as the "Contract Sum") over the Price for Work Done to Date (referred to in other contracts as the "Gross Valuation").

So, if the direct fee percentage was 8 per cent and the Price for Work Done to Date at termination is £2,500,000 and total of the Prices at the Contract Sum was £4,200,000, the amount payable by the Employer to the Contractor is £136,000:
(£4,200,000 – £2,500,000) x 8% = £136,000

Interest on late payment

Note that with late or non-payments from the Employer to the Contractor, under Clause 51.2 of the Engineering and Construction Contract, the Contractor is entitled to be paid interest which is assessed from the date by which the late payment should have been made until the date when the payment is made. This interest is calculated on a daily basis at the interest rate Contract Data Part 1 and is compounded annually.

Option Y(UK)2: The Housing Grants, Construction and Regeneration Act 1996

Note that under Option Y(UK)2 (if selected), which relates to the Housing Grants, Construction and Regeneration Act 1996 and the Local Democracy, Economic Development and Construction Act 2009, the Contractor may exercise his right under the Act to suspend performance due to late or non-payment.

It is also a compensation event, so the Contractor can recover any Defined Cost or delay to Completion as a result of exercising his right to suspend all or part of the works.

Chapter 10

Dealing with disputes

Adjudication and tribunal

Question 10.1 What is the difference between "Option W1" and "Option W2" within the NEC3 Engineering and Construction Contract?

The NEC3 Engineering and Construction Contract contains two dispute resolution procedures: Option W1 and W2.

- Option W1 is used unless the Housing Grants, Construction and Regeneration Act 1996 applies.
- Option W2 is used when the Housing Grants, Construction and Regeneration Act 1996 applies.

Note: The Local Democracy, Economic Development and Construction Act 2009, which came into force in England and Wales on 1 October 2011 and in Scotland on 1 November 2011, has amended the Housing Grants, Construction and Regeneration Act 1996, the new Act applying to most UK contracts after that date.

Option W1

Within Option W1 there is an Adjudication Table under which, if the referring party is the Contractor and the dispute is due to an action or inaction of the Project Manager or Supervisor, he must notify the dispute to the Employer within four weeks of becoming aware of the action or inaction, and then refer it to the Adjudicator between two and four weeks after that notification.

If the referring party is the Employer and the dispute is regarding a compensation event which has been treated as having been accepted (Clause 62.6), then the Employer must notify the dispute to the

Contractor and then refer it to the Adjudicator between two and four weeks after that notification.

This seems an odd reason for referring a dispute to the Adjudicator as Clause 62.6 states that if the Project Manager does not reply to the Contractor's notification it is treated as acceptance of the quotation. How could an Adjudicator whose role is to enforce the contract find other than in favour of the Contractor?

Some practitioners have stated that this inclusion could in theory allow the Employer to override the actions of his own Project Manager, but the referral to the Adjudicator is initiated by the Project Manager notifying the dispute to the Employer and the Contractor, so the Employer could not initiate the action himself.

For any other matter the Employer wishes to refer to the Adjudicator, it may be referred between two and four weeks after notifying the Contractor and the Project Manager.

Whilst these timescales are fixed within the Adjudication Table, they can be extended by the Project Manager if the Contractor and the Project Manager agree before either the notice or the referral is due.

Note that if the matter in dispute is not notified and referred within these timescales, neither Party can refer the dispute to the Adjudicator or the tribunal.

The Adjudicator is appointed under the NEC3 Adjudicator's Contract. He acts impartially and, if he resigns or is unable to act, the Parties jointly appoint a new Adjudicator. One would assume that the referring party obtains a copy and completes the Adjudicator's Contract.

If the parties have not appointed an Adjudicator, either party may ask the nominating body to choose an Adjudicator within four days of the request. No procedures have been specified for appointing a suitable person, and in practice a number of different methods have been used.

Whatever method is used, it is important that both Parties have full confidence in his impartiality, and for that reason it is preferable that a joint appointment is made. The Adjudicator should be a person with experience in the type of work included in the contract between the Parties and who occupies or has occupied a senior position dealing with disputes. He should be able to understand the viewpoint of both Parties.

Often the Parties delay selecting an Adjudicator until a dispute has arisen, although this frequently results in a disagreement over who should be the Adjudicator. As noted, the selection of the Adjudicator is important, and it should be recognised that a failure to agree an Adjudicator means that a third party will make the selection without necessarily consulting the Parties.

The referring party must include within his referral information that he wishes to be considered by the Adjudicator. Any more information from either party to be provided within four weeks of the referral.

A dispute under a subcontract, which is also a dispute under the Engineering and Construction Contract, may be referred to the Adjudicator at the same time, and the Adjudicator can decide the two disputes together.

The Adjudicator may review and revise any action or inaction of the Project Manager, take the initiative in ascertaining the facts, and the law relating to the dispute, instruct a party to provide further information and instruct a party to take any other action which he considers necessary to reach his decision.

All communications between a party and the Adjudicator must be communicated to the other party at the same time.

The Adjudicator decides the dispute and notifies the parties and the Project Manager within four weeks of the end of the period for providing information. This four-week period may be extended by joint agreement between the parties.

Until this decision has been communicated, the parties proceed as if the matter in dispute was not disputed.

The Adjudicator's decision is binding on the parties unless and until revised by a tribunal and is enforceable as a contractual obligation on the parties.

The Adjudicator's decision is final and binding if neither party has notified the Adjudicator that they are dissatisfied with an Adjudicator's decision within the time stated in the contract, and intends to refer the matter to the tribunal.

The Adjudicator may, within two weeks of giving his decision to the parties, correct a clerical mistake or ambiguity.

Review by the tribunal

A dispute cannot be referred to the tribunal unless it has first been referred to the Adjudicator.

The tribunal may be named by the Employer within Contract Data Part 1.

Whilst no alternatives are stated, it will normally be litigation or arbitration. If arbitration is chosen the Employer must also state in Contract Data Part 1 the procedure, the place where the arbitration is to be held, and the person or organisation who will choose an Arbitrator if the Parties cannot agree a choice, or if the named procedure does not state who selects the Arbitrator.

A party can, following the adjudication, notify the other party within four weeks of the Adjudicator's decision that he is dissatisfied. This is a time-barred right, as the dissatisfied Party cannot refer the dispute to the tribunal unless it is notified within four weeks of the Adjudicator's decision; failure to do so will make the Adjudicator's decision final and binding. Also, if the Adjudicator has not notified his decision within the time provided by the contract, a party may within four weeks of when the Adjudicator should have given his decisions notify the other party that he intends to refer the dispute to the tribunal.

The tribunal settles the dispute and has the power to reconsider any decision of the Adjudicator and review and revise any action or inaction of the Project Manager or Supervisor. It is important to note that the tribunal is not a direct appeal against the Adjudicator's decision; the parties have the opportunity to present further information or evidence that was not originally presented to the Adjudicator, and also the Adjudicator cannot be called as a witness.

Option W2

The inclusion of adjudication within the original New Engineering Contract, now the NEC3, pre-dates UK legislation giving parties to a contract the statutory right to refer a dispute to adjudication "at any time", the key difference between Options W1 and W2 being that Option W2 does not have fixed timescales for notification of a dispute to the other Party and also to the Project Manager.

The Housing Grants, Construction and Regeneration Act 1996 The Local Democracy, Economic Development and Construction Act 2009 which came into force in England and Wales on 1 October 2011 and in Scotland on 1 November 2011 amends the Housing Grants, Construction and Regeneration Act 1996, the new Act applying to most UK contracts after that date.

There still remains certain categories of contract to which neither Act applies. A new Secondary Option has been introduced to all NEC3 contracts to cover the provisions of the new Act.

The differences between the Housing Grants, Construction and Regeneration Act 1996 and the Local Democracy, Economic Development and Construction Act 2009 have been explored in a previous Chapter but in respect of adjudication they include the following:

- The previous Act only applied to contracts which were in writing; the new Act also applies to oral contracts.
- Terms in contracts such as "the fees and expenses of the Adjudicator as well as the reasonable expenses of the other party shall be the responsibility of the party making the reference to the Adjudicator" will be prohibited. Much has been written about the effectiveness of such clauses, often referred to as Tolent clauses (Bridgeway Construction Ltd v Tolent Construction Ltd 2000) and whether they comply with the previous Act, so the new Act should provide clarity for the future.
- The Adjudicator is permitted to correct his decision so as to remove a clerical or typographical error arising by accident or omission. Previously he could not make this correction.

Section 108 of the Act provides parties to construction contracts with a right to refer disputes arising under the contract to adjudication. It sets out certain minimum procedural requirements which enable either party to a dispute to refer the matter to an independent party who is then required to make a decision within 28 days of the matter being referred.

If a construction contract does not comply with the requirements of the Act, or if the contract does not include an adjudication procedure, a statutory default scheme, called the Scheme for Construction Contracts (referred to as the "Scheme"), will apply.

The Act provides that a dispute can be referred to adjudication "at any time" provided the parties have a contract. This is not necessarily during the construction stage, for example a designer may refer a dispute during the design stage. "At any time" can also refer to a dispute after the contract is completed.

What are the requirements?

Section 108 of the Act requires all "construction contracts", as defined by the Act, to include minimum procedural requirements which enable the parties to a contract to give notice at any time of an intention to refer a dispute to an Adjudicator. The contract must provide a timetable so that the Adjudicator can be appointed, and the dispute referred, within seven days of the notice.

The Adjudicator is required to reach a decision within 28 days of the referral, plus any agreed extension, and must act impartially. In reaching a decision an Adjudicator has wide powers to take the initiative to ascertain the facts and law related to the dispute. For this option there is no Adjudication Table, as the parties have a right to refer a dispute to each other and to the Adjudicator at any time.

The Housing Grants, Construction and Regeneration Act 1996 defines adjudication as: a summary non-judicial dispute resolution procedure that leads to a decision by an independent person that is, unless otherwise agreed, binding upon the parties for the duration of the contract, but which may subsequently be reviewed by means of arbitration, litigation or by agreement. In that sense, adjudication does not necessarily achieve final settlement of a dispute because either of the parties has the right to have the same dispute heard afresh in court, or where the contract specifies arbitration.

However, experience since the Housing Grants, Construction and Regeneration Act 1996 came into force shows that the majority of adjudication decisions are accepted by the parties as the final result.

The Adjudicator is appointed under the NEC Adjudicator's Contract; he acts impartially, and if he resigns or is unable to act, the Parties jointly appoint a new Adjudicator.

If the parties have not appointed an Adjudicator, either party may ask the nominating body to choose an Adjudicator within four days of the request. A party may first give a notice of adjudication to the other party with a brief description of the dispute, details of where and when the dispute has arisen, and the nature of the redress sought.

The Adjudicator may be named in the contract in which case the party sends a copy of the notice to the Adjudicator. The Adjudicator must confirm within three days of receipt of the notice that he is able to decide the dispute, or if he is unable to decide the dispute.

Within seven days of the issue of the notice of adjudication, the party:

- refers the dispute to the Adjudicator
- provides the Adjudicator with the information on which he relies together with supporting information
- provides a copy of the information and supporting documents to the other party.

Again, a dispute under a subcontract, which is also a dispute under the Engineering and Construction Contract, may be referred to the Adjudicator at the same time and the Adjudicator can decide the two disputes together.

The Adjudicator may review and revise any action or inaction of the Project Manager, take the initiative in ascertaining the facts, and the law relating to the dispute, instruct a party to provide further information, and instruct a party to take any other action he considers necessary to reach his decision.

All communications between a party and the Adjudicator must be communicated to the other party at the same time. The Adjudicator decides the dispute and notifies the parties and the Project Manager within four weeks of the end of the period for providing information. This four week period may be extended by joint agreement between the parties.

Until this decision has been communicated, the parties proceed as if the matter in dispute was not disputed.

The Adjudicator's decision is binding on the parties unless and until revised by a tribunal and is enforceable as a contractual obligation on the parties.

The Adjudicator's decision is final and binding if neither party has notified the Adjudicator that they are dissatisfied with an Adjudicator's decision within the time stated in the contract, and intends to refer the matter to the tribunal.

The Adjudicator may, within two weeks of giving his decision to the parties, correct any clerical mistake or ambiguity.

Question 10.2 When are we required to select and name the Adjudicator on our NEC3 Engineering and Construction Contract?

Within Contract Data Part 1 there are three options for selecting and naming the Adjudicator.

1 The name and address of the Adjudicator may be stated by the Employer in Contract Data Part 1 and he is then appointed before the starting date. This has the advantage that tendering Contractors are aware at the time of tender who will be the Adjudicator in the event that a dispute arises and, if necessary, can object to them within their tenders. Also, the Adjudicator is already in place should a dispute arise. The named Adjudicator may require some form of retainer fee for being named in the contract and being available in the event that a dispute arises.
2 The parties can mutually agree to the name of the Adjudicator in the event that a dispute arises. This has the advantage that there is no "Adjudicator in waiting" and therefore no fee payable. Many Engineering and Construction Contract practitioners say that

pre-appointing the Adjudicator as in Option 1 is resigning oneself to the fact that there will at some point be a dispute. However, once parties are in dispute they then have to agree who is to be the Adjudicator and appoint him. At this point, though, they may not wish to agree with each other about anything!

3 The name of the Adjudicator nominating body may be stated by the Employer in Contract Data Part 1. This is probably the favoured option, as an independent name can be put forward by the nominating body, who is normally well prepared to put forward the name of an Adjudicator and have him appointed within the timescales set by the contract and if necessary the appropriate legislation.

In all cases, the Adjudicator must be impartial, i.e. the Adjudicator should not show any bias towards either party. In addition, all correspondence from the Adjudicator must be circulated to both parties. Any request for an Adjudicator must be accompanied by a copy of the notice of adjudication, and the appointment of the Adjudicator should take place within seven days of the submission of the notice of adjudication to the other party.

Adjudicator Nominating Bodies, as the name suggests, are organisations that fulfil the role of nominating Adjudicators. These bodies keep registers of Adjudicators with varying expertise and based at various geographical locations who can act for parties should they be nominated.

Below is a list of recognised Adjudicator nominating bodies:

- Association of Independent Construction Adjudicators (AICA)
- Centre for Effective Dispute Resolution (CEDR)
- Chartered Institute of Arbitrators (CIArb)
- Chartered Institute of Arbitrators (Scotland) (CIArb-Scotland)
- Chartered Institute of Building (CIOB)
- Construction Conciliation Group (CCG)
- Construction Confederation (CC)
- Construction Industry Council (CIC)
- Construction Plant-Hire Association (CPA)
- Dispute Board Federation (DBF)
- Dispute Resolution Board Foundation (DRBF)
- Institution of Chemical Engineers (IChemE)
- Institution of Civil Engineers (ICE)
- Institution of Electrical Engineers (IEE)

- Institution of Mechanical Engineers (IMechE)
- Law Society of Scotland (LawSoc (Scot))
- Nationwide Academy of Dispute Resolution (NADR)
- RICS-Dispute Resolution Service (RICS-DRS)
- RICS Dispute Resolution Service Australia (RICS DRS (Oceania))
- Royal Incorporation of Architects in Scotland (RIAS)
- Royal Institute of British Architects (RIBA)
- Royal Institution of Chartered Surveyors (RICS)
- Royal Society of Ulster Architects (RSUA)
- Technology and Construction Court Bar Association (TECBAR)
- Technology and Construction Solicitors Association (TeCSA)

Clearly, in naming the Adjudicator Nominating Body in the contract it is advisable to name an organisation with expertise in the project to be carried out. These bodies are very knowledgeable about appointment of Adjudicators and the relevant timescales and, for a modest fee, can nominate a suitably qualified Adjudicator to suit the parties' requirements.

Chapter 11

Preparing and assessing tenders

Completing Works Information, Site Information and Contract Data, inviting tenders, etc.

Question 11.1 What documents do we need to compile to invite tenders for an NEC3 Engineering and Construction Contract using Option B (priced contract with bill of quantities)?

Whilst any list of contract documents cannot be exhaustive as it depends on the type of project, the scope of works, requirements of the Employer, etc., there are certain documents that will be required as a minimum when inviting tenders for an Engineering and Construction Contract using Option B.

These will normally consist of the following:

(i) Invitation to Tender (including instructions to tenderers on time and place of tender submission)

There is no "pro forma" version of "Instructions to Tenderers" within the NEC contracts, but any Invitation to Tender will give a brief description of the work, who it is for, and how tenders may be submitted.

Other inclusions will cover such matters as:

- time, date and place for the delivery of tenders
- what documents must be included with the tender, including a programme
- the policy regarding alternative and/or non-compliant bids
- arrangements for visiting the Site including contact details
- rules on non-compliant bids
- anti-collusion certificate.

(ii) Form of Tender

There is no "pro forma" Form of Tender within the NEC contracts, though the Engineering and Construction Contract Guidance Notes include a sample form. This is the tenderer's written offer to execute the work in accordance with the tender documents.

The Form of Tender is normally in the form of a letter with blank spaces for tenderers to insert their name and other particulars, total tender price, and other particulars of their offer. It is essential to have a standard Form of Tender and that all tenderers consistently use the same form, to make comparison of tenders easier.

(iii) The Pricing Document.

In this case it will be a Bill of Quantities prepared in accordance with the relevant Standard Method of Measurement as defined within Contract Data Part 1.

(iv) Works Information

Works Information is defined by Clause 11.2(19) as "information which either specifies and describes the works or states any constraints on how the Contractor Provides the Works and is either in the documents which the Contract Data states it is in, or in an instruction given in accordance with this contract".

The Employer provides the Works Information and refers to it in Contract Data Part 1, if the Contractor provides Works Information, e.g. for his design, this is included in Contract Data Part 2.

The main documents within the Works Information are normally the drawings and specifications, but will also include those listed below.

Description of the works

- A statement describing the scope of the works.
- Schematic layouts, plan, elevation and section drawings, detailed working and/or production drawings (if relevant and available), etc.
- A statement of any constraints on how the Contractor Provides the Works, e.g. restrictions on access, sequencing or phasing of works, security issues, etc.

Plant and Materials

- Materials and workmanship specifications.
- Requirements for delivery and storage.
- Future provision of spares, maintenance requirements, etc.

Health and safety

- Specific health and safety requirements for the Site which the Contractor must comply with, particularly if the Site is within existing premises, including house and local safety rules, evacuation procedures, etc.
- Any pre-construction information and health and safety plans for the project.

Financial Records

- Details of the accounts and records to be kept by the Contractor.

Contractor's design

- The default position is that all design will be carried out by the Employer.

If any or all of the design is to be carried out by the Contractor, it should be included within the Works Information together with any performance requirements to be met within the Contractor's design. In addition, any warranty requirements and also any future novation requirements should be included within the Works Information.

Any design acceptance procedures should be included including timescales for submission and acceptance of design.

Completion

Completion is defined under Clause 11.2(2), but the Works Information should also define any specific requirements required in order for completion to have taken place for example, and requirement for:

- "as built" drawings.
- maintenance manuals.
- training documentation.

- test certificates.
- statutory requirements and/or certification.

Services

- Details of other Contractors and Others who will be occupying the Working Areas during the contract period and any sharing requirements.

Subcontracting

- Lists of acceptable subcontractors for specific tasks.
- Statement of any work which should not be subcontracted.
- Statement of any work which is required to be subcontracted.

Programme

- Any information which the Contractor is required to include in the programme in addition to the information shown in Clause 31.2. Also, if the Employer requires the Contractor to produce a certain type of programme or to submit it using a certain brand of software, then this should be clearly detailed within the Works Information.

Tests

- Description of tests to be carried out by the Contractor, the Supervisor and Others, including those which must be done before Completion.
- Specification of materials, facilities and samples to be provided by the Contractor and the Employer for tests.
- Specification of Plant and Materials to be inspected or tested before delivery to the Working Areas.
- Definition of tests of Plant and Materials outside the Working Areas which have to be passed before marking by the Supervisor.

Title

- Statement of any materials arising from excavation or demolition to which the Contractor will have title (Clause 73.2).
- Requirements for Plant and Material to be marked (Clause 71.1).

Others

There are also certain specific requirements for statements to be made in the Works Information from certain Main and Secondary Options in the conditions of contract.

- X4.1 – the form of any parent company guarantee.
- X14.2 – the form of any advance payment bond.
- X13.1 – the form of any performance bond.

The Works Information must be carefully drafted in order to define clearly what is expected of the Contractor in the performance of the contract and therefore included in the quoted tender amount and programme. If the contract does not cover all aspects of the work, either specifically or by implication, that aspect may be deemed to be excluded from the contract.

A comprehensive, all-embracing description should therefore be considered for the scope of work clause, which should be supplemented by specific detailed requirements.

If reliance is placed solely on a very detailed scope description, an item may be missed from this detailed description and be the subject of later contention.

Where items of equipment are to be fabricated or manufactured off Site by others, it is advisable that the contract sets out the corresponding obligations and liabilities of the respective parties, particularly if these are to form an integral or key part of the completed works.

The Works Information describes clear boundaries for the work to be undertaken by the Contractor. It may also outline the Employer's objectives and explain why the work is being undertaken and how it is intended to be used. It says what is to be done (and maybe what is not included) in general terms, but not how to do it or the standards to be achieved. It explains the limits, where the work is to interface with other existing or proposed facilities. It may draw attention to any work or materials to be provided by the Employer or others. It should also emphasise any unusual features of the work or contract, which tenderers might otherwise overlook.

This is the document that a tenderer can look to, to gain a broad understanding of the scale and complexity of the job and be able to judge its capacity to undertake it. It is written specifically for each contract. In some respects it is analogous to a shopping list. It should be comprehensive, but it should be made clear that it is not intended to include all the detail, which is contained in the drawings, specifications and schedules.

The Works Information will also include drawings which again should provide clear details of what the Contractor has to do. Clearly, tenderers must be given sufficient information to enable them to understand what is required and thus submit considered and accurate tenders.

It will also include the specification, which is a written technical description of the standards and various criteria required for the work, and should complement the drawings. The specification describes the character and quality of materials and workmanship for work to be executed.

Again, it may lay down the order in which various portions of the work are to be executed. As far as possible, it should describe the outcomes required, rather than how to achieve them. It is customary to divide the work into discrete sections or trades (e.g. drainage, concrete, pavements, fencing, etc.) with clauses written to cover the materials to be used, the packaging, handling and storage of materials (only if necessary), the method of work to be used (only if necessary), installation criteria, the standards or tests to be satisfied, any specific requirements for completion, etc.

The specification is an integral part of the design. This is often overlooked, with the result that inappropriate or outmoded specifications are selected, or replaced by a few brief notes on the Drawings. The designer should spend an appropriate amount of time specifying the quality of the work, as it is not possible to price, build, test or measure the work correctly unless this is done.

The Works Information describes what the Contractor has to do in terms of scope and standards, and in some cases must not do or include in order to Provide the Works. It may also include the order or sequence in which the works are to be carried out. It also details where the work is to interface with other existing or proposed Contractors or facilities. It will also include any work or facilities or materials to be provided by the Employer or others. Additionally, it may include any unusual features of the work that tenderers might otherwise overlook, for example any planning constraints, etc.

This is the document from which a tenderer can gain a broad understanding of the scale and complexity of the job and his capacity to undertake it. It is written specifically for each contract.

The Works Information should be:

- clear – unambiguous
- concise – not excessively wordy
- complete – have nothing missing.

In respect of the Contractor's design, a reason for not accepting the design is that it does not comply with either the Works Information or the applicable law. Again, if the Works Information was lacking and the Contractor provided a design that complied with it, and the applicable law, could the Contractor escape liability for the defective design?

Secondary Option X15 states that "the Contractor is not liable for Defects in the works due to his design so far as he proves that he used reasonable skill and care to ensure that his design complied with the Works Information". Again, this places a heavy burden on the Works Information.

(v) Site Information

Site Information is defined in Clause 11.2(16) as "information which describes the Site and its surroundings" and is identified in Contract Data Part 1.

Site Information may include the following:

- Ground investigations, borehole and trial pit records and test results. If the Employer has obtained such information, it should not be withheld from the tenderers, though the tenderers should be aware that the Site Information alone cannot be relied upon in terms of a possible compensation event (see Clause 60.2).
- Information about existing buildings, structures and plant on or adjacent to the Site.
- Details of any previously demolished structures and the likelihood of any residual surface and subsurface materials.
- Reports obtained by the Employer concerning the physical conditions within the Site or its surroundings. This may include mapping, hydrographical and hydrological information.
- All available information on the topography of the Site should be made accessible to tenderers, preferably by being shown on the Drawings.
- Environmental issues, for example nesting birds and protected species.

- References to publicly available information.
- Information from utilities companies and historic records regarding plant, pipes, cables and other services below the surface of the Site.

It is vital that care is taken to get the Site Information correct. Tenderers must be given sufficient information to enable them to understand what is required and thus to submit considered and well-priced tenders.

Under Clause 60.3, if there is an ambiguity or inconsistency within the Site Information, the Contractor is assumed to have taken into account the conditions most favourable to doing the work. Whilst many interpret that as the Contractor allowing the cheapest way of doing the work, it may be the easiest or quickest way.

In the event of the Contractor notifying a compensation event under Clause 60.1(12) for unforeseen physical conditions, he is assumed to have taken into account the Site Information, publicly available information referred to in the Site Information, information obtained from a visual inspection of the Site, and other information which an experienced contractor could reasonably be expected to have or to obtain. So the Contractor cannot rely solely on the Site Information in terms of how he prices and programmes the works.

(vi) Contract Data Part 1 (completed by the Employer)

The Contract Data provides the information required by the conditions of contract specific to a particular contract. Other contracts call this the "Appendices" or "Contract Particulars".

Part 1 consists of data provided by the Employer, the sections of the Contract Data aligning with the sections of the core clauses.

Section 1 – General

This section requires the Employer, or the party representing the Employer, to identify the selected Main and Secondary Options, a description or title for the works, and the names of the Employer, the Project Manager, the Supervisor and the Adjudicator.

Whilst the Project Manager and the Supervisor must always be named individuals, many Employers choose to insert company names for these parties within the Contract Data and to separately identify the named individuals. It is also not uncommon for Employers to name directors or partners of their respective companies, then the authority of the Project Manager and Supervisor is delegated to the individuals under Clause 14.2.

The Works Information and Site Information are also identified within this section, though these are normally incorporated by reference to separate documents rather than listing drawing numbers and specification references. If this is the case, the separate documents must be clearly defined.

The boundaries of the Site are defined, normally by reference to a specific drawing or map.

As the NEC3 contracts are intended for use worldwide, the language and the law of the contract are also entered.

The "period for reply" is the period that the parties have to reply to submissions, proposals, notifications, etc. where there is no period specifically stated elsewhere within the contract – for example, the period within which a party should reply to an early warning notice – or the Project Manager should reply to the Contractor submitting the particulars of his design or the name of a proposed Subcontractor.

The "period for reply" would clearly not apply to the Contractor's submission of a programme for acceptance or the submission of a quotation for a compensation event.

With regard to dispute resolution, the Adjudicator nominating body is named within this section, normally a professional institution, and then the tribunal is named as either arbitration or legal proceedings.

Finally within this section, any matters to be included in the Risk Register will be identified.

Section 2 – Contractor's main responsibilities

This section is not included within the Contract Data as no entries are required.

Section 3 – Time

The starting date, access dates, and the frequency of requirement for the Contractor to submit revised programmes are inserted within this section. Many Employers change the reference to "one calendar month" rather than "weeks", to align with monthly progress meetings or reporting requirements.

Section 4 – Testing and Defects

The "defects date" is identified as a number of weeks after Completion of the whole of the works. This identifies the period the Contractor is initially liable for correcting Defects.

The defect correction period is also stated, this being the period within which the Contractor must correct each notified Defect, failing which the Project Manager assesses the cost to the Employer of having the Defect corrected by Others.

The Contract Data provides for three entries to be inserted here if required, which then allows for different types of Defect to have different correction periods dependent on the urgency to have them corrected, for example:

- mechanical and electrical works – 24 hour
- drainage – 48 hours
- finishings – 7 days.

Clearly, any defect that has a potential effect on health and safety should be corrected as soon as possible.

Section 5 – Payment

Again, as the NEC3 contracts are intended for use worldwide, the currency of the contract is entered.

The assessment interval is also entered, which again is often expressed as "one calendar month" rather than in "weeks". The Project Manager decides the first assessment date to suit the parties and following assessment are carried out within the assessment intervals.

Section 6 – Compensation events

The only entries within this section are in respect of the weather, the place where weather is to be recorded (weather station, airport, etc.), the weather measurements to be recorded, the default measurements being cumulative monthly rainfall, the number of days with rainfall more than 5 mm, minimum air temperature less than 0 degrees Celsius, and snow lying at the designated time of day. The supplier of the weather measurements, the place where weather measurements are recorded, and where they are available from are also stated.

For some isolated site locations where no recorded data may be available, assumed values may be inserted.

Section 7 – Title

This section is not included within the Contract Data as no entries are required.

Section 8 – Risks and insurance

The minimum limit of indemnity for third party public liability and the Contractor's liability for his own employees is stated within this section.

Optional statements

Contract Data Part 1 contains a series of optional statements which are completed where appropriate.

First, if the tribunal has been identified in Section 1 as arbitration, the Employer must identify the arbitration procedure, the place where any arbitration is to be held, who will choose an arbitrator if either the parties cannot agree a choice, or if the arbitration procedure does not state who selects an arbitrator.

If the Employer has decided the completion date for the whole of the works, the date is inserted in this section; alternatively, the tendering Contractors may be required to insert a date in Contract Data Part 2.

When the Contractor completes the works, the Project Manager certifies Completion and the Employer takes over the works not later than two weeks after Completion, even though take over may be before the Completion Date. If the Employer is not willing to take over the works before the Completion Date, the statement to that effect should remain in the Contract Data, if not it should be deleted. This statement is referred to in Clause 35.1.

The timing for submission of the Contractor's first programme is stated as a period of weeks from the Contract Date, the date when the contract came into existence.

If the Employer has identified work which is to meet a Key Date (Clause 11.2(9)), this is stated by the Employer as a condition to be met, and the Key Date.

If Option Y(UK)2 is not used, the period for payment is three weeks from the assessment date (Clause 51.2), though this can be amended in this section.

Similarly, if Option Y(UK)2 is used, the final date for payment is 14 days after the date when the payment is due, and again this can be amended within the section.

If there are additional Employer's risks (Clause 80.1) these can be listed.

There are then optional statements regarding insurance.

First, if the Employer is to provide plant and materials there is provision for insurance of the works to include any loss or damage of such plant and materials.

The Contractor provides the insurances stated within the Insurance Table, except any insurances the Employer is to provide that are stated in Contract Data Part 1, which lists what the insurance is to cover, the amount of cover and the deductibles (otherwise known as "excesses") – the amounts to be paid by the insuring party, often before the insurance company pays.

Dependent on which Main Option is selected, the Employer enters details for the relevant entry in the Contract Data:

- If Options B or D are used, the Employer states the method of measurement and any amendments.
- If Options C or D are used, the share range and Contractor's share percentage are stated.
- If Options C, D, E or F are used, the intervals within which the Contractor prepares forecasts of Defined Cost is defined. Also, the exchange rates to be used and the date on which they are published.

Finally, dependent on which Secondary Options are selected, the Employer enters details for the relevant entry in the Contract Data.

(vii) Contract Data Part 2 (blank pro forma to be completed by tendering Contractors)

Part 2 is completed by the Contractor and includes the names of key people and documents submitted with the tender, and also various percentages for fee, people overheads, Working Areas overheads, design overheads, manufacture and fabrication overheads, and adjustment for listed Equipment, etc., appropriate to the choice of main option. It is critical that these percentages are analysed as part of the tender assessment process.

In addition, there may be other information, dependent on the applicable legislation – for example, in the UK, where the Construction (Design and Management) Regulations 2007 apply, Pre-Construction Information would be included within the tender documents.

Question 11.2 We are preparing the Bill of Quantities for a project to be carried out under Option B (priced contract with bill of quantities). Is the preparation of the Bill of Quantities any different from any other contract? The Employer wishes to make a decision about certain landscaping elements as the project proceeds, so we are intending to include Provisional Sums within the Bill of Quantities for the time being.

First, the preparation of a Bill of Quantities for an NEC3 Engineering and Construction Contract is the same as for any other contract, so the same Rules of Measurement will be used as defined within Contract Data Part 1 and the quantities would be entered into the Bill of Quantities in the normal way.

However, one would have to give some thought as to the normal "Preambles" and "Preliminaries" sections as they would need to be labelled as "Works Information" (information which "specifies and describes the works or states any constraints on how the Contractor Provides the Works") and "Site Information" ('information which describes the Site and its surroundings").

With regard to the question about using Provisional Sums for certain landscaping elements, let us first consider the definition of Provisional Sums.

Provisional Sums are used where there are elements of work which are not designed or cannot be sufficiently defined at the time of tender and therefore a sum of money is included by the Employer in the Bill of Quantities or other pricing document to cover the item.

When the item is defined or able to be properly defined, the Contractor is given the information which allows him to price it; the Provisional Sum is omitted and the price included in its stead.

The problem with Provisional Sums is that they reduce the competition amongst tenderers as they are not priced at tender stage, and also if they are "defined" the Contractor is deemed to have allowed time in his programme for them, if they are "undefined" he has not.

The NEC3 contracts do not provide for Provisional Sums, the principle being that:

1 the Employer decides what he wants at tender stage so it can be designed and accurately described and the tendering Contractors can properly price and programme for them, or
2 when the Employer decides what he wants the Project Manager can give an instruction to the Contractor which changes the Works Information; it is a compensation event under Clause 60.1(1) and can be priced at the time. In that case, the cost of the work is held in the scheme rather than in the contract.

Index